PLANT SCIENCE RESEARCH AND PRACTICES

PHYTOCHEMICALS

PLANT SOURCES AND POTENTIAL HEALTH BENEFITS

PLANT SCIENCE RESEARCH AND PRACTICES

Additional books and e-books in this series can be found on Nova's website under the Series tab.

PLANT SCIENCE RESEARCH AND PRACTICES

PHYTOCHEMICALS

PLANT SOURCES AND POTENTIAL HEALTH BENEFITS

IMAN RYAN
EDITOR

nova
science publishers
New York

Library of Congress Cataloging-in-Publication Data

ISBN: 978-1-53615-478-8

Published by Nova Science Publishers, Inc. † New York

CONTENTS

PREFACE

The opening chapter of *Phytochemicals: Plant Sources and Potential Health Benefits* discusses macronutrients and micronutrients from plants along with their benefits to human health.

An overview of the mechanisms of action underlying the potential roles of dietary soybean isoflavones including daidzein, genistein, and equol that may have benefits in dyslipidemia, ischemic heart disease, blood pressure, diabetes, cancer, and osteoporosis is discussed in the next chapter.

In one study, the authors seek to scientifically validate the hypoglycaemic effects of crude aqueous extracts from Vitex payos fruits on both normal and alloxan-induced diabetic rats.

Additionally, the authors discuss the literature results obtained in the identification and quantification of rosmarinic acid and its derivatives in *Lamiaceae* family plants, as well as explore recent advances in the nanoencapsulation of rosmarinic acid.

This book goes on to report several aspects of the chemical composition and biological activity of species of *Smilax* found in Brazil.

Smilax fluminensis is a dioecious liana with a great capacity for shoot formation in soil and primary branch formation in subsoil. As such, phytochemical studies on this plant are reviewed.

Phytochemical aspects and biological activities of the species *S. brasiliensis* are also examined.

To continue the discussion, the authors investigate the allelopathic, cytotoxic, genotoxic and larvicidal potential of methanol and petroleum ether extracts from *S. brasiliensis* leaves, as well as the effects of their fatty acids and methyl esters.

Furthermore, the authors synthesize phytochemical aspects and biological activity of *Tecoma stans* (L.) species.

In the penultimate chapter, the authors discuss *P. venusta*, and the way in which compounds obtained from this plant can be used as functional foods, drugs or precursors for the semi-synthesis of drugs.

The objective of the concluding chapter is to present the ethnopharmacological, biological activities and phytochemical aspects of the *Limoniastrum feei* species growing in the Algerian Sahara.

Chapter 1 - The chapter aims to bring an up to date about macronutrients and micronutrients from plants along with their benefits for human health. It is very well known that in order to survive plants require a minimum of 14 mineral elements for the nutrition that are generally absorbed from the soil through a plant's roots. The chapter aims to: i) present the main role in the human body of the macronutrients found in plants referring to nitrogen, phosphorus, potassium, calcium, magnesium, sulphur as well as micronutrients found in plants pointing towards chlorine, iron, zinc, boron, copper, manganese, nickel, molybdenum, selenium, iodine; ii) a second aim would be to discuss each of the aforementioned macronutrient and micronutrient in terms of availability in medicinal plants, in order to be able to quantify the correlation between there commended daily dose and the consummation of medicinal plants or standardized extracts.

Chapter 2 - Soybeans are among the most popular foods worldwide and consumption of soy-containing foods has been associated with health benefits. Soybeans contain saponins, soy proteins and peptides, and flavonoids, all of which exert multiple biological activities. In particular, numerous epidemiologic studies have indicated an inverse relationship between the consumption of soy isoflavones and the incidence of lifestyle-related diseases. For example, soy isoflavones reduce plasma total cholesterol and triglycerides and increase plasma leptin levels. In addition, consumption of isoflavones is associated with a lower risk of type 2 diabetes in post-menopausal women. Epidemiological studies also indicate that high consumption of soy isoflavones is associated with a lower incidence of hormone-dependent cancers, such as breast and prostate cancer. In ovariectomized mice, isoflavone-enriched dietary supplementation prevents bone mineral density reduction and enhances bone formation. However, although the evidence supporting the benefits of soy phytochemicals in disease prevention continues to increase, the association between soy isoflavones and disease is not fully understood. Typical soy isoflavones include daidzein and genistein. Furthermore, equol, a major metabolite of daidzein, is generated by enterobacterial effects. This review provides an overview of the recent progress in identifying the mechanisms of action underlying the potential roles of dietary soybean isoflavones including daidzein, genistein, and equol that may have benefits in dyslipidemia, ischemic heart disease, blood pressure, diabetes, cancer, and osteoporosis.

Chapter 3 - This study sought to scientifically validate the hypoglycaemic effects of crude aqueous extracts from *Vitex payos* fruits on both normal and alloxan-induced diabetic rats. Powdered pulp of *V. payos* fruit was hydrated with distilled water to 0.27 g/ml and water soluble compounds were extracted. Alloxan-induced diabetic rats were used to determine the hypoglycaemic effect of the extract after 16 days of treatment in comparison to distilled water and glibenclamide controls. Glucose tolerance test was also done to determine the effectiveness of the plant extracts as a hypoglycaemic agent. Alloxan-induced diabetic rats in the negative control group given distilled water died before the second measurement of blood glucose on the third day of treatment. Over the 16 day period, a general decrease in blood glucose levels to 5.2 ± 0.3 mmol/L, was observed in rats that were treated with glibenclamide. In the alloxan-induced diabetic groups that were administered with crude aqueous extracts of *V. payos*, a dose dependent decrease in blood glucose levels was observed. The rats that were administered with 800mg/kg of extract had their glucose levels decreasing more than those that were administered with doses of 400mg/kg and 100mg/kg. There was no significant difference ($P > 0.05$) in blood glucose levels of normal rats that were administered with different concentrations of the crude extracts of *V. payos* fruits compared to the normal rats that were given distilled water (control group) over the 16 days of treatment. After initial glucose feeding in the OGTT test, an increase in blood glucose levels was observed in both normal and alloxan-induced diabetic rats that were administered with extracts of *V. payos*. After 60 minutes all the groups administered with extracts of *V. payos* had glucose levels gradually decreasing. In the OGTT test normal rats that were given distilled water (negative control group) had their glucose levels increasing until the end of 120 minutes. Rats that were treated with glibenclamide had an initial increase in glucose level over the first 30 minutes of glucose feeding after which glucose levels began to decrease. Crude aqueous extracts from fruits of *V. payos* showed effective hypoglycaemic activity in the alloxan-induced diabetic rats and could potentially work as a prophylactic in deterring hyperglycaemia.

Chapter 4 - Rosmarinic acid (RA) was first isolated from *Rosmarinus officinalis* L., since then it has been identified in many plants within the *Lamiaceae* family, such as species belonging to the genus *Rosmarinus*, *Thymus*, *Salvia* or *Origanum*, among others. Its biosynthesis begins from L-phenylalanine and L-tyrosine amino acids, resulting in an ester of caffeic acid. Rosmarinic acid, among the phenolic acids, is one of the most studied. *In vitro* and *in vivo* bioactivities of RA have been studied and are of interest to several fields of knowledge related to human health. RA and its derivatives offer a wide range of scientifically proven bioactivities, such as anti-oxidant, anti-inflammatory, anti-angiogenic, anti-proliferative, anti-microbial and others. In this review the authors aim to: i) discuss the literature results obtained in identification/quantification of RA and its derivatives in *Lamiaceae* family plants; ii) to explore the bioactivities of RA and derivatives, such as anti-inflammatory, antioxidant, anti-proliferative and anti-microbial, that are well described in various cell lines and microorganisms, with a diversity of assays, correlating the methods of analysis and the structure-dependent activities between RA and its derivatives. Beyond the health effects of RA, this chapter also aims to explore the most recent advances in the nanoencapsulation of RA focusing on human health applications. RA is chemically unstable and shows poor solubility in water which constrains its transport across biological barriers. Its loading in nanoformulations has been anticipated to preserve the phenolic acid bioactivities and to enhance its bioavailability.

Chapter 5 - Brazil stands out as one of the most biodiverse countries on the planet. It is estimated that Brazilian ecosystems house about 20% of all biological species in the world. The medicinal plants of Brazil have stood out for their importance in popular use. It is widely documented and proven that medicinal plants are a source of unpublished molecules. The family *Smilacaceae* belongs to the group of monocotyledons, possessing 310 species distributed in temperate, tropical and subtropical regions. The family consists of dioecious species, vines or herbaceous vines, and may occur as sub-shrubs or shrubs, with simple, alternating, petiolate leaves with reticulated veins. *Smilax* is the largest genus in the *Smilacaceae* family and is represented in Brazil by 33 species widely distributed throughout the country, but mainly in the Cerrado. Plants of this genus are known as "sarsaparilla" and "japecanga" and are used in folk medicine as a tonic, antirheumatic, anti-syphilitic, antihypertensive and diuretic and for sweating, arthritis and skin conditions. The genus *Smilax* has proved to be a source of secondary metabolites that have several types of biological activity. Phytochemical investigations have shown the presence of polar compounds such as: saponins and a complex mixture of phenolic compounds such as flavonoids, caffeic acid, *p*-coumaric acid, ferulic acid and *trans*-cinnamic acid in the species present in Brazil. Regarding the importance of studies of species of the genus *Smilax*, few studies are available in the literature. The biological activity already reported for Brazilian species of this genus is antioxidant, antifungal, antibacterial, antinociceptive, anti-inflammatory and allelopathic in nature. The objective of this review is to report aspects of the chemical composition and biological activity of the species of the genus *Smilax* found in Brazil.

Chapter 6 - *Smilax fluminensis* is popularly known as japecanga or sarsaparilla. This species is a dioecious liana with a great capacity for shoot formation in the soil and primary branch formation in the subsoil. It has a wide geographic distribution, occurring over a range of environments from closed areas such as the Amazon Forest to open places such as the Pantanal. The adaptive success of the species can be attributed to its genotypic plasticity. Although the genus *Smilax* is not well studied, the use of *S. fluminensis* has been reported in folk medicine since antiquity, with antibacterial, antifungal, anti-inflammatory and hematoprotective activities, as well as in the treatment of syphilis and rheumatism. Here the authors aim to review the first phytochemical studies on this plant. Phytochemical work using the leaves of *S. fluminensis* indicated the presence of flavonoids, isolating and identifying quercetin-3-*O*-α-rhamnopyranoside (1-6)-*O*-β-D-glucopyranoside and quercetin-3-*O*-β-D-galactopyranoside. In another study, the presence of the following compounds was detected: coumarins, tannins, flavonoids, steroids, steroidal saponins glycosides, cardiotonic glycosides, and alkaloids. Chlorogenic acid, caffeic acid, rutin, *p*-coumaric acid, and *trans*-cinnamic acid were identified by high performance liquid chromatography (HPLC). The first study showing the facility for propagation of this species, through the use of seeds and remains of seedlings and shoots, corroborates its potential economic use in a sustainable way, constituting an alternative to reducing the predatory extraction of this native species, if its pharmacological potential is confirmed.

Chapter 7 - *Smilax brasiliensis* Sprengel is a monocotyledon belonging to the family *Smilacaceae*. It is a native species of the Brazilian Cerrado and presents itself as a shrub, a climbing vine, or even a tree, being popularly known as japecanga or sarsaparilla. In folk medicine, it is used for the treatment of sexually transmitted diseases, cutaneous conditions, and inflammatory diseases and as a diuretic, diaphoretic, and depurative agent, the roots, and rhizomes being the main parts used. This review will report on the phytochemical aspects and biological activities of the species *S. brasiliensis*. Phytochemical investigations have already been carried out with different parts of the plant, such as the leaves, roots, rhizomes, stems, and branches. Studies with extracts and fractions of this species have demonstrated the presence of phenolic compounds, such as rutin, quercetin, and chlorogenic acid, and apolar compounds, especially fatty acids, with a high proportion of linolenic, linoleic, and palmitic acids. Other classes of secondary metabolites responsible for the diversity of biological actions in the plant have also been identified, among them coumarins, alkaloids, flavonoids, condensed tannins, steroids, triterpenoids, essential oils, and saponins, the latter being the main class of secondary metabolites found in the *Smilax* genus. Saponins are compounds with a steroidal structure and are of great industrial interest, since they have applicability in the production of anti-inflammatory drugs and contraceptives. As *S. brasiliensis* is still a little-studied species, only a few reports of its biological activities were found, such as antibacterial, antifungal, antioxidant, allelopathic, antihyperglycemic, antihyper-lipidemic, and anticancer effects. No toxic effects on *Artemia salina* were observed, and it was not genotoxic. Considering that this species is not well explored with regard to its biological actions and the chemical compounds present in the same, *S. brasiliensis* can be an alternative source of pharmacological activities for application in the improvement of health and the environment.

Chapter 8 - Many commercially available synthetic compounds are used as pesticides and herbicides; however, many of these may be toxic to man and to the environment. Plants may be an alternative source of compounds with similar properties, but without such toxicity. *Smilax brasiliensis* Sprengel is a medicinal species from the Brazilian Cerrado that is used to treat syphilis and skin disorders in folk medicine. This study aimed to investigate the allelopathic, cytotoxic, genotoxic and larvicidal potential of methanol and petroleum ether extracts from *S. brasiliensis* leaves, as well as the effects of their fatty acids and methyl esters. Extracts, fatty acids and methyl esters inhibited the growth of *Allium cepa* radicles and hypocotyls, and pure substances principally inhibited the growth of radicles. For *Lactuca sativa*, samples from *S. brasiliensis* leaves and pure substances showed diverse allelopathic activity with regard to radicle and hypocotyl growth. The frequency of chromosomal abnormalities was significant without a dose-dependent effect. *S. brasiliensis* did not show action on the cell cycle of *A. cepa*. The tested samples showed no larvicidal activity against *Culex quinquefasciatus* larvae. Additional studies should evaluate the possibility of using *S. brasiliensis* extracts, fatty acids and methyl esters as natural herbicides.

Chapter 9 - Nowadays there is increasing interest in medicinal plants as well as the use of herbal medicines to treat various pathologies. The species *Tecoma stans* (L.) Kunth belongs to the *Bignoniaceae* family which is composed of 860 species, grouped in 100 predominantly Neotropical genera. *T. stans* is a tree native to Mexico and the southern United States, having been introduced in Brazil as an ornamental plant. It occurs as a cultivated plant or of sub-spontaneous origin in almost the entire territory. This literature review synthesizes phytochemical aspects recently elucidated and biological activity for *T. stans* species. Phytochemical investigations have revealed the presence of saponins, flavonoids, monoterpene alkaloids, tannins, phenolic compounds, steroids, anthraquinones, terpenes, hydrocarbons, volatile oils, glycosides, carbohydrates, proteins, and resins in *T. stans* extracts. Throughout the world, several types of biological activity have been reported for the species, including antimicrobial, antioxidant, antidiabetic, antitumor, antinociceptive, anti-inflammatory, and cytotoxic activity. However, as it is considered exotic in Brazil and is used on a large scale for ornamentation, studies of the medicinal potential of *T. stans* in the country are still rare. The variety of biological activity together with the phytochemical composition of *T. stans* suggest that the species represents an important source of medicinal properties and has great potential for improvement of the quality of life and welfare of the population, evidencing the need for and relevance of more studies of this species, especially in Brazil.

Chapter 10 - Many plant species produce bioactive compounds, which can play an important role in human health. *Pyrostegia venusta* (Ker Gawl.) Miers (synonyms: *Pyrostegia ignea* and *Bignonia venusta*) belongs to the *Bignoniaceae* family and is a wood-climbing vine dispersed in tropical regions. *Bignoniaceae* Juss. is represented by about 82 genera and 800 tree species, shrubs and creepers, with pantropical distribution and pronounced occurrence in the neotropics. The family is subdivided into eight tribes which are differentiated by plant habit and fruit morphology. Brazil has some endemic taxa and is considered the center of diversity of the family, with 32 genera and 350 species occurring in several types of environment from the Brazilian Cerrado to perennial wet forests. *P. venusta* is a seasonal species easily recognized by the abundant production of orange flowers. In folk medicine, *P. venusta* is used to treat coughs, bronchitis, infections, erysipelas, jaundice, vitiligo and diarrhea. Ethnopharmacological surveys consistently describe its potential as a medicinal plant, and much pharmacological activity is related to phenolic compounds, mainly flavonoids, such as hesperidin, homoeriodictyol, rutin, quercetin, catechin and acacetin-7-*O*-glycoside found in leaves. Other bioactive compounds are found in flowers and roots, such as carotenoids, β-amyrin, β-sitosterol, *myo*-inositol, oleanolic acid, hexadecanoic acid, linoleic acid, oleic acid and allantoin. Recent studies showed an absence of genotoxic activity in the ethanolic extracts of *P. venusta* flowers, and good ability of methanolic extracts of leaves, flowers and roots to scavenge free radicals. Other biological activity has been reported, such as the improvement of flu symptoms, stimulation of melanogenesis, wound healing and anti-inflammatory, antinociceptive and antimicrobial activity. *P. venusta* is an interesting source of chemiodiversity, and the compounds obtained from this plant can be used as functional foods, drugs or precursors for the semi-synthesis of drugs. So, this plant represents an important source of bioactive compounds of pharmaceutical/medicinal interest, with recognized biological potential.

Chapter 11 - Traditional remedies have been employed for the treatment and management of various ailments since the beginning of human civilization. The plant family *Plumbaginaceae* comprises a large number of species (836 species grouped in 27 genera) that have been and are still used as medicinal plants, particularly in folk medicine. In the Algerian Sahara, the *Plumbaginaceae* family is represented only by two genera *Limonium* and *Limoniastrum.* The medicinal specie *Limoniastrum feei* (Girard) Batt., is endemic to Algeria and Morocco growing in arid conditions has been used ethnobotanically for treatment of a variety of disorders, bronchitis, diarrhea, gastrointestinal tracts, as anti-inflammatory, for skin diseases, treatment of infected wounds, hepatic pains, children fever and in addition the plant provides good grazing for camel. The aim of this chapter is to present as much information as was established from the available scientific literature. The review comprises the ethnopharmacological, biological activities and phytochemical of the *Limoniastrum feei* specie growing in the Algerian Sahara and used as medicinal plants.

In: Phytochemicals
Editor: Iman Ryan
ISBN: 978-1-53615-478-8

Chapter 1

ESSENTIAL MINERAL ELEMENTS: MACRONUTRIENTS AND MICRONUTRIENTS FROM HERBS IN HUMAN HEALTH

Corina Danciu[1], Diana Antal[1], Stefana Avram[1], Valentina Buda[2,*], Ioana Zinuca Pavel[1], Daliana Minda[1], Florina Ardelean[1], Mirela Nicolov[3] and Cristina Dehelean[4]

[1]Department of Pharmacognosy, "Victor Babeş" University of Medicine and Pharmacy, Timişoara, România

[2]Department of Pharmacology and Clinical Pharmacy, "Victor Babeş"University of Medicine and Pharmacy, Timişoara, România

[3]Department of Pharmaceutical Physics, "Victor Babeş" University of Medicine and Pharmacy, Timişoara, România

[4]Department of Toxicology, "Victor Babeş" University of Medicine and Pharmacy, Timişoara, România

* Corresponding Author's E-mail: buda.valentina.oana@gmail.com.

ABSTRACT

The chapter aims to bring an up to date about macronutrients and micronutrients from plants along with their benefits for human health. It is very well known that in order to survive plants require a minimum of 14 mineral elements for the nutrition that are generally absorbed from the soil through a plant's roots. The chapter aims to: i) present the main role in the human body of the macronutrients found in plants referring to nitrogen, phosphorus, potassium, calcium, magnesium, sulphur as well as micronutrients found in plants pointing towards chlorine, iron, zinc, boron, copper, manganese, nickel, molybdenum, selenium, iodine; ii) a second aim would be to discuss each of the aforementioned macronutrient and micronutrient in terms of availability in medicinal plants, in order to be able to quantify the correlation between there commended daily dose and the consummation of medicinal plants or standardized extracts.

Keywords: macronutrients, micronutrients, plants, benefits, human health

INTRODUCTION

All living organisms require inorganic elements for their life processes. A number of minerals are though essential in various vital biochemical processes. Plants, animals and humans have certain mineral requirements for homeostasis and healthy functioning, while deficiencies or excess may be responsible for important health issues [1].

It is unanimously accepted that in order to grow and develop plants need essential mineral elements. According to the needed amount in the plant, these minerals are divided into macroelements and microelements [2]. Moreover it is well known that the concentrations of macroelements and microelements may vary from one species to another and may also oscillate within the same species depending on pedoclimatic conditions. The minimum amount required for plant growth have been previously described by Epstein and have the following concentration within the plant (mg/kg-weight of dry matter): nitrogen (15,000), phosphorus (2,000), potassium (10,000), calcium (5,000), magnesium (2,000), sulphur (1,000), chlorine (100), iron (100), zinc (20), boron(20), copper(6), manganese

(50), nickel (0.1), molybdenum (0.1) [3]. As a classification rule, the elements which are necessary in high amounts are called macroelements and are usually involved in the building of the skeleton of the plant or different protoplasmic constituents, while microelements also called trace elements are required in small amounts and usually participate in various metabolism pathways. Microelements haven't got an important role in the development of osmotic potential while macroelements may have. On the other hand, only a slight excess of microelements can be toxic for the plant, while accumulation of macroelements do not cause such phenomena [4].

Water and soil are the major sources of essential minerals for plants, while human intake depend on their diet, including the consumed plants. Mineral imbalances or deficiencies represent serious public health issues in many parts of the world, especially for vulnerable categories as pregnant women and children [1]. Nutrition is an important source for essential minerals and health care. The available information on essential minerals and plant intake is still poor. Our material intends to contribute to the field, with relevant data regarding the amounts of minerals in certain type of plants and plant-foods, vital for balanced diets and mineral supplementation.

Main Macronutrients That Can Be Found in Plants

Nitrogen

Nitrogen is an important element in the structure of nucleic acids and amino acids, therefore it is necessary for the plant in high amounts [5]. It is taken from the environment as nitrates and its assimilation depends on the enzyme nitrate reductase. This enzyme reduces NO_3^- to NO_2^-, followed by the reduction of NO_2^- to ammonium, which is assimilated in chloroplasts. Plant growth and development depends on nitrogen metabolism [6]. In plants, nitrogen is present both in the structure of proteins as well as in secondary metabolites: alkaloids, cyanogenic glycosides and glucosinolates [7].

Some plants, especially those belonging to the *Fabaceae* family, establish symbiosis with nitrogen-fixing bacteria such as rhizobia. The bacteria provide the plant with nitrogen and it receives in return carbohydrates synthesized during photosynthesis [8]. This mutual advantageous symbiosis causes the formation of nodules in the roots [9]. Plants in the *Fabaceae* family produce a higher nitrogen containing metabolites than other plants, and these products are usually stored in seeds [7]. The seeds of these plants contain between 17 and 40% proteins and are an important source of nitrogen from diet [10].

Besides *Fabaceae*, seeds of plants from several other families are considered protein sources. *Chenopodium quinoa* (*Amaranthaceae*) seeds contain ten essential amino acids: lysine, methionine, tyrosine, leucine, isoleucine, tryptophan, phenylalanine, histidine, valine, threonine. The protein content is between 12.5 and 16.7 g/100 g fresh weight [11]. Chia (*Salvia hispanica - Lamiaceae*) seeds contain 16-23% proteins [12] and amaranth (*Amaranthus*sp.) also has a higher protein content than other plants, ranging from 13 to 19%. The proteins from amaranth seeds are a good source of lysine, tryptophan and sulphur containing amino acids [13]. The nutritional value of quinoa, chia and amaranth seeds made them appreciated sources of proteins with a balanced content of essential amino acids [14].

Phosphorus

Phosphorus is found in the constitution of phospholipids, nucleic acids and several coenzymes. Its important role in the plant is evidenced by a decrease in carbon fixation as a consequence of phosphorus deficiency [15]. In plants, it has a role in respiration, photosynthesis, as well as in the synthesis of nucleic acids. Plants contain up to 0.5% phosphorus [16]. The assessment of total phosphorus content in different medicinal plants revealed a higher content in *Urtica doica* leaves, compared to *Achillea millefolium*, *Betula pendula* and *Hypericum perforatum* [17].

Phosphorus from plants has a low bioavailability because it is commonly stored as phytic acid or phytates, compounds that can't be degraded due to the absence of enzyme phytase in humans. Nuts, cereals or beans contain phytic acid and phytates [18]. Phytate content is 9.2-16.7 mg/g in soybeans, 10.6-15.1 mg/g in amaranth grain and 9.2-16.2 mg/g in buckwheat. The phytates from diet have a negative impact on minerals uptake, especially calcium, iron, magnesium and zinc. They form insoluble complexes that reduce the availability of minerals [19].

Potassium

Potassium is the main cation in plants, having important functions in the osmotic regulation of the cellular protoplasm, stomatal movement, enzyme activation, nitrogen uptake and the synthesis of proteins [20]. Its content in plants is genetically regulated, but influenced by the available amount from the soil, the age of the plant, the species and the plant organ. Plants which store increased amounts of complex sugars and proteins have increased K demands [21]. Higher K concentrations in aerial parts are correlated to the role of potassium in photosynthesis. Potassium is as well a key element in stress response [22].

The potassium content of various plant products collected from wild-growing species in situated between 1-6g/kg dry mass: birch *(Betulapendula)* bark (1.1g/kg); alder *(Alnusglutinosa)* bark (1.7g/kg); spruce (*Piceaabies*) bark (1.9g/kg); raspberry *(Rubusidaeus)* twigs (1.9g/kg); young oak (*Quercusrobur)* twigs (2.0g/kg); blueberry (*Vacciniummyrtillus*) twigs (2.4g/kg); beechnuts (*Fagussylvatica*) (4.8g/kg), sessile oak (*Quercuspetrea*)glands (6.4g/kg). Vegetables and fruits currently used in human nutrition have comparatively higher potassium contents, related to the use of K-rich fertilizers: oranges (17.9 g K/kg dry mass); strawberries (21.6g/kg); tomatoes (48.9g/kg); potatoes

(26)g/kg; cabbage (26.7-66.1g/kg); broccoli (75.4g/kg); salad (78.2g/kg); spinach (82.0g/kg) [23].

A research performed on 56 wild-growing medicinal plants revealed a K content ranging from 4.2 g/kg in ash (*Fraxinus excelsior*) leaves to 64.7 g/kg in the herb of greater celadine (*Chelidonium majus*) [24]. Among them, a number of herbs could be pointed out as particularly rich in potassium, while in the meantime devoid of toxic secondary metabolites: mallow (*Malvasylvestris*) leaves and flowers (40.2 g/kg),elder *(Sambucus nigra*) flowers (38.9g/kg), wild pansy (*Viola tricolor*) herb (36.8 g/kg), dandelion (*Taraxaci herba*) herb (34.8 g/kg), wild garlic (*Alliumursinum*) herb (29.2 g/kg), stinging nettle (*Urtica dioica*) leaves (24.2 g/kg),horsetail (*Equisetum arvense*) herb (23.5 g/kg), goldenrod (*Solidagovirgaurea*) herba (22.2 g/kg) and raspberry (*Rubusidaeus*) leaves (20.6g/kg) [25]. These natural products can be employed as aqueous extracts, salad additives or encapsulated powder. Taking into account the essentiality of potassium for human physiology and the fact that approximately 10% of the population has a marginal potassium deficiency [21], it is important to insure an adequate of the element. In addition, the presence of potassium promotes, in synergy with secondary metabolites like flavonoids and saponins, a diuretic-aquaretic effect based on osmotic mechanisms [26]. Of particular interest in this context is the K/Na ratio which should be above 100:1; such examples include nettle leaves (458:1), mallow leaves and flowers (428:1) elder flowers (339:1); horsetail (281:1), and Saint John's wort (238:1) [24].

Calcium

Calcium is an essential element for both plants and humans. In plants, Ca is a structural element of the cell wall, where it balances the negative charge of pectins, but controls as well wall enzymes. It regulates ion exchange in the cell wall and membrane, and acts as a couter-ion for anions in the vacuole. Moreover, Ca^{2+} is a key secondary messenger of physiologic processes involved in development and response to environmental challenges [27]. Most of calcium is present in complexed form: oxalate, fiber, phytate, proteins and other compounds, limiting its bioavailability for humans from plant-based nutrients [28]. Calcium oxalate makes up between 3-80% of a plant's dry mass, and sequesters up to 90% of the total calcium [29]. The bioavailability of Ca from plants is inversely proportional to oxalate content; a typical example in this regard is spinach [30]. The highest Ca amounts are concentrated in leaves [31].In fruits, Ca stabilizes the cell membranes and delays ripening; mature fruits contain low Ca amounts. The Ca content of various plant nutrients ranges from0.8-1.5 g Ca/kg dry mass in cereals, 0.29 in potatoes, 4.5 in cauliflower, 5.8 in carrots, 10.0 in cabbage, 26.7 in spinach. The bioavailability of Ca from plants is inversely proportional to oxalate content; a typical example in this regard is spinach. According to Fleming and Heimbach [32], plants representing good Ca sources must offer a minimum of 30 mg absorbable Ca per serving, while providing less than 100kcal.

Among medicinal plants, several species contain high Ca amounts. Nettle leaves have a particularly high Ca content (36.4 g/kg dry mass) while being rich in many other essential macro- and microelements [33]. Ash leaves are as well Ca-rich (35 g/kg), being followed by horsetail (26g/kg), ribwort plantain (22g/kg), wood garlic (15g/kg), linden flowers with bracts (15g/kg). Considering a normative intake of 500 mg Ca/day [34], ten gram nettle leaves may cover up to 70% of this intake. The extraction of the element through decoction varies from 17.5% (elder flowers) to 77.7% (restharrow roots) [24]; in infusions the extraction with water ranges from 10.6 for lavender to 71.6% for coriander fruits, contributing with about 1% of the recommended daily intake [35].

Magnesium

Magnesium is a highly important cation in plants. It is a building element of chlorophyll, a cofactor of many enzymes, involved in ATP formation in chloroplasts, carbon dioxide fixation during photosynthesis, utilization of photo assimilates, protein synthesis and other functions [36,

37]. In humans, it affects several metabolic pathways involving ATP utilization, energy production and storage, transmission of nerve impulse, muscle contraction, cardiac excitability, blood pressure, and other [38]. Plant-based foodstuffs are a major Mg source, and countries which rely more on vegetable foods are less at risk for Mg deficiency as Western countries where meat, dairy products and processes foods are important elements of the diet [39]. Green vegetables and whole grains have a particularly high Mg content [40]. Some examples of Mg content in plants are as follows: spinach (5.0 g/kg dry mass); lettuce (3.4g/kg), cabbage (1.8g/kg); cauliflower (2.1g/kg); tomatoes (1.7g/kg); onions (1.3g/kg);potatoes (1.3g/kg); kiwis (1.2g/kg); wheat (1.2g/kg), rye (1g/kg), apples (0.4g/kg), cherries (0.7g/kg) [41, 42].

Medicinal plants are rich in magnesium, especially aerial (green) parts have high Mg content correlated with their chlorophyll content. Examples in this regard are: chamomile (3.15 g/kg), green tea (2.1 g/kg) [43], ash leaves (5.5g/kg); elder flowers (5.5 g/kg);meadowsweet (5.3g/kg); wild pansy herb (4.7g/kg); raspberry leaves (4.5g/kg); horsetail (4.2g/kg); mallow leaves and flowers (4g/kg); stinging nettle leaves (3.9g/kg); wild thyme (3.2g/kg); hawthorn flowers with leaves (3.1g/kg) [24]; lavender (4.4g/kg); mint (7.2g/kg); plantain (3.9g/kg); dill (3.8g/kg); coriander (3.2g/kg); echinacea (1.7g/kg) [44]. Magnesium has a good extraction rate in water. For infusions, determined percentages are 47–48% from chamomile [45], 21-37% from chamomile, mint, St John's wort, nettle and sage [35]; for decoctions extraction percentages range from 50% for meadowsweet and elder, to 80% for wild pansy [24]. Hence, herbal teas may represent a useful Mg supplement in human diets. For instance, the intake of a cup of mint tea with a content of 37.5 mg Mg/L, offers 2.5% of the recommended dietary intake for Mg [35].

Sulphur

Sulphur is found in plants in the structure of proteins, vitamins such as vitamin B1 or biotin and in the structure of some secondary metabolites.

The amino acids cysteine and methionine from plants are important sources of sulphur containing amino acids for humans [46]. Plants take up sulphur from the soil mainly as sulphate. The sulphate is used for the production of cysteine [47] that will be further used for the synthesis of other compounds [46]. Sulphur content in plants is between 0.1 and 1.5% dry weight [48].

Sulphur containing secondary metabolites show important therapeutic properties. The glucosinolates, found mainly in plants from the *Brassicaceae* family, are secondary metabolites that after hydrolysis by the enzyme myrosinase form products with antioxidant, anti-inflammatory, antibacterial or anticancer properties [49]. *Allium* species present sulfoxides and glutamyl peptides as main sulphur compounds [50]. Allicin from *Allium sativum* (garlic) and *Allium ursinum* (ramsons) is known as an antimicrobial, lipid-lowering and anti-cancer agent. It is obtained from alliin after hydrolysis by the enzyme alliinase [51].

Although sulphur is important for the proper functioning of human metabolism, a recommended daily intake has not been established. Even though most of the sulphur comes from the intake of amino acids cysteine and methionine, the sulphur metabolites produced by alliaceous and cruciferous plants also have a contribution [52]. However, the recommended dietary allowance is indicated only for sulphur containing amino acids (14 mg/kg body weight) [46, 52].

MAIN MICRONUTRIENTS THAT CAN BE FOUND IN PLANTS

Chlorine

Chlorine mainly found in plants as the anion chloride (Cl^-), represents an essential microelement for several metabolic processes of plants. Especially accumulating in chloroplasts, chloride plays an important role in photosynthesis. Chloride is also known as a catalyst for several enzymes

[53]. Moreover, chloride is essential for stomatal regulation, contributing to its normal opened and closure, being also involved in osmotic regulation. Different studies indicated different types of interaction between chloride and other minerals [54].In human, chloride is the most important anion in the extracellular fluid, involved in electrolyte balance, and the major anion required for gastric secretion [55].

Sources of chloride are volcanic rocks, water and salt. Generally, soils provide high concentration of chloride, so that plants become also a source, with an average content 2-20g/kg dry matter, much above plant requirements (0.2-0.4 g/kg dry matter) [54]. Also the concentrations found in drinking water and salt, usually are sufficient to fulfill the recommended daily intake (3400 mg/day) [56].Most of the chloride is distributed in plants in the vegetative organs as shoots and leaves, and less than 1% in seeds. Some plants have higher requirements, therefore higher concentrations were signaled in kiwi fruit (2.13 g/kg dry matter), sugar beet (0.71 g/kg dry matter), and palm trees such as coconut and oil palm (2.49 g/kg dry matter) [54].

Medicinal plants are not regularly investigated concerning their content in chlorine. The amount of this essential microelement is usually high in plants, surpassing the required concentrations. Available data about chlorine concentration in some medicinal plants collected from uncontaminated areas in Romania, showed the following: coriander 553.02 μg/g dry matter, lavender 996.93μg/g dry matter, mint 3731.17 μg/g dry matter, echinacea 3897.67 μg/g dry matter, plantain 4080 μg/g dry matter, dill 5747 μg/g dry matter, chamomile 8210.21μg/g dry matter [35]. Plants and medicinal plants in particular remain a minor source for recommended essential chlorine in human nutrition.

Iron

Iron is an essential element for both plants and vertebrates. In plants, it is essential for vital biochemical processes such as mitochondrial respiration, chlorophyll formation, as cofactor in protein regulating processes. Soil is the primary source of iron for plants, but mostly present in complexed forms with low water solubility, thus hardly bioavailable [57]. Yet, too much iron can as well become toxic for living cells. Organisms have developed specific mechanisms in order to regulate iron homeostasis. The phenomenon is linked to human iron intake from diet, and is correlated to the iron deficiency considered one of the 20 important risk for worldwide diseases burden [58]. Although non-heme iron contained by plants is inferior and much more variable compared to the heme-iron type, there are several strategies, especially for low income countries, or for those embracing a vegetarian diet, to diminish iron deficiencies that include plant fortification. Iron is frequently added in agricultural fertilizers [59], or directly fortifying food products with more available iron forms. Recent advances in the fields of iron homeostasis in plants are being materialized in biofortification with accessible forms of iron especially for staple plants like whole cereals (e.g., maize, rice, wheat, millet) or pulses (e.g., beans, peas, lentils) [59]. Transgenic varieties of some crops, increased concentrations from 15 μg/g dry matter (rice), 25 μg/g dry matter (wheat) to values as high as100 μg/g dry matter, with 2-6 x fold increase of iron content compared to regular crops [60]. Other vegetal rich sources in iron, that are commonly recommended for vegetarian diets include cashew nuts and sesame seeds (approximately 50 μg/g), or pumpkin seeds (approximately 100 μg/g) [61]. Whereas among green vegetables the iron amounts found in spinach vary from approximately 350 μg/g dry matter [62] and up 10 times lower for the raw product [61]. Other leafy vegetables such as rucola contain around 11μg/g fresh weight, watercress 9 μg/g fresh weight or kale 4 μg/g fresh weight are also sources of iron but concentrations are lower compared to spinach. Some more concentrated sources of iron are represented by algae; species like *Sargassum* with 156 μg/g dry matter, *Gracilariopsis* with 195μg/g dry matter and *Ulva* with 57 μg/g dry matter, if rationally used in diets could contribute to iron deficiencies in underdeveloped countries [63].

Other sources of inorganic and organic type of iron are food supplements, which contain more and more plants or plant extracts. Even if inorganic herbal iron has a lower bioavailability, iron concentration in medicinal species of interest, next to phyto compounds that are known as absobtion inhibitors, remain still important. A semi-systematic review indicates that iron is especially concentrated in roots [64].Another study that investigated a number of medicinal plants from Pakistan, found some species, like *Menthaarvensis, Glycyrrhizaglabra, Rheum emodi*, with very high levels of iron (above 4000 μg/g dry matter) [65]. The level of iron in some investigated medicinal plants from Romania, showed values ranging from48 μg/g dry matter for coriander and 52 μg/g dry matter for echinacea to high iron concentrations in dill (134 μg/g dry matter), chamomile (239 μg/g dry matter), mint (498 μg/g dry matter) to very high yields in plantain (1409 μg/g dry matter) and lavender (3315 μg/g dry matter) [35]. Others analyzed the content in iron in some edible and medicinal plants and concentrations ranging from 17 μg/g fresh product for *Plantago major* to 39 μg/g fresh product for *Verbena officinalis* [66].

Zinc

Zinc is an essential microelement in all living organisms, as cofactor to many enzymes involved in vital processes as cell replication, gene expression and protein metabolism. Zinc deficiency leads in plants to impaired development, disease vulnerability, sterility and harvest quality [67]. Humans also depend on zinc for basic cell functions, enzymatic processes and the well function of the immune system. Zinc deficiencies are reported for more than 30% of people around the world [1] and contribute with 1.9% to the worldwide burden of disease [68]. Low intakes, next to absorption issues are the main causes of Zn deficiency. Some of the most important sources of this element are cereal husks and nuts, but also meat, eggs, seafoods and dairy products [1]. As in the case of iron, the bioavailability of Zn from vegetarian sources is inferior to the non vegetarian ones, due to high phytic acid content, an absorption inhibitor. High phytate products like whole cereals are also higher in zinc compared to refined foods, but still, vegetarian diets are associated to reduced zinc concentrations. Beside supplementation, dietary diversification may represent a sustainable strategy to reduce micronutrient deficiency Zinc concentration in edible largely consumed plants are therefore of importance both for vegetarians but also for possible supplementation from vegetal sources [69].

A large surface of the cultivated soils are low in zinc and there are strategies for zinc and calcium bioavailability, such as some agricultural processes affecting root absorption, fertilization with easily available forms of zinc, genetic biofortification which can generate concentrations at least 2-fold higher than regular crops [70]. Targeted concentrations for zinc in edible crops were set to 28 μg/g dry matter in rice, 38μg/g dry matter in wheat and maize, 56 μg/g dry matter in beans, 66 μg/g dry matter in pearl millet, 70 μg/g dry matter in sweet potatoes. Zinc concentrations up to 500-5000μg/g dry matter in roots and 100-700 μg/g dry matter in leaves can be reached, while in fruits, seeds and tubers, concentrations higher than 100 μg/g dry matter are possible only after biofortification to improve phloem translocation [68].

Beside edible plants, medicinal plants are also rich in essential minerals, including zinc, and can contribute to the mineral requirements in humans, due to more and more frequent use of such remedies. Several published papers report the content of zinc in medicinal plants from various areas in the world, indicating a great variability. In plants from Brazil, the majority of results indicated values up to approximately 50 μg/g dry matter, while high concentrations of zinc were found in few species like *Salvia officinalis* (102 μg/g dry matter), *Equisetum arvense* (169 μg/g dry matter) [71]. In a set of plants from Pakistan the range of zinc was from approximately 17μg/g dry matter in *Convulvulus arvensis* to 65 μg/g dry matter in *Hordeum vulgare* [72]. Less data is available on the content of minerals in plant extracts. A study investigated some plant extracts showing yield about 20-50% for zinc after hot water extraction, with higher values in birch leave extract [73]. Zinc content from some leafy greens in India, reported values from around 30 μg/g dry matter in amaranth to 50 μg/g dry matter in cauliflower and mint and 100 μg/g dry matter in spinach [62]. Some medicinal plants from Romania showed values from 8 μg/g dry matter in echinacea to as high as 55 μg/g dry matter in coriander [34].

Boron

Boron is a trace mineral considered as a micronutrient for plants, animals and humans. It is mainly found in forms of organic complexes such as boric acid and borate [74]. Though less concentrated in nature is commonly present in rocks, soil and water. In plants, boron is known to be involved in various fundamental biological function, as nucleic acid, protein, carbohydrate synthesis, cell function and integrity, phenol metabolism [75]. Humans benefit from boron as a microelement in important processes, among others, bone homeostasis, wound healing and inflammatory processes, hormone and enzyme balance, antioxidative processes [74]. Recommended doses are not to be surpassed, developmental and reproductive functions being affected. Plants represent a source of boron, as its absorption from the soil is regulated by the plants in even low supplies. In some edible plants boron levels vary on a range from 10 μg/g dry matter in broccoli to 96 μg/g dry matter in tomatoes, with intermediate values for potatoes 21050 μg/g dry matter and corn 15-90 μg/g dry matter [76].Some of the most concentrate plant-foods in boron are considered avocados with 14 μg/g product, 5.9 μg/g in peanut butter, 5.8 9 μg/g in dry peanuts, 5.6 9 μg/g in prune juice, 4.3 9 μg/g in cacao powder.

Medicinal plants absorb boron, hence representing an alternative source. Different types of *Camellia sinensis* leaves were investigated, indicating amounts ranging from around 3to 17 μg/g dry matter [77]. Less data is available regarding boron content in medicinal plants. A study measured the content of some herbal mixtures recommended in diabetes, containing boron and reported concentrations of approximately 45 μg/g dry matter for *Myrtilli folium, Phaseolifructus sine seminibus* and 53 μg/g dry matter for *Myrtilli folium, Phaseolifructus sine seminibus* and *Salviae folium* [78].

Copper

Copper is an essential trace metal, usually existing as monovalent and divalent cation, being involved in several functions in plants, such as in photosynthesis, mitochondrial function, hormonal signaling or as enzyme cofactor, along with zinc [79]. Deficiencies occur in case of low availability, affecting growth and reproduction. Copper toxicity is also important, when high concentrations are reached, impairing several vital functions. In humans, Cu is essential for the hematologic, neurologic and bone metabolism, being involved also in iron transfer to plasma, while toxic concentrations cause impaired liver function [1].Important sources of copper include liver, whole grains, legumes, nuts, seafoods [1].Some other plants represent good sources of copper, such as bean, black pepper, Brazilian nut, cashew, cocoa, prunes [77]. In a study investigating edible plants, higher values were reported for *Amaranth viridis* with 2.2 μg/g fresh and for *Verbena officinalis* with 3.3 μg/g [66]. Absorption from vegetal sources has to consider the interaction between Cu and Zn in terms of competitive absorption [67].

Medicinal plants are also sources for copper. For instance, various brands of *Camellia sinensis* teas were investigated and values were higher for black teas, largely varied up to 602 μg/g dry matter [77]. Another study found that among the analyzed species, *Hordeum vulgare* (approximately 19 μg/g dry matter), *Ricinus comunnis* (approximately 15 μg/g dry matter) showed highest copper concentrations [72]. Medicinal plants from Romania were investigated, indicating values ranging from 15 μg/g dry matter for echinacea, to approximately 31 μg/g dry matter for lavender [34].

Selenium

Selenium is an essential element for humans that possesses anti-inflammatory, antioxidant and antiviral effects [80]. For plants it is a non-essential mineral element and it is known that in excessive amounts it can be toxic [81, 82]. It can be found in various plants and its concentration is dependent to selenium levels present in the soils. Characteristics of the soil such as pH, microbial activity and redox status can impact selenium assimilation by plants [83]. For humans, selenium has also an important role in cancer prevention and in the immune system, stimulating antibody formation [83]. In the environment, selenium can be found both in the organic (selenomethionine – Semet and selenocysteine - Secys) or the inorganic forms (selenite, selenide, selenate and also as selenium element). After plants decomposition, the organic forms are the main components, whereas the inorganic forms are usually present in most soils. The inorganic forms, selenite and selenite are commonly present in the aerial parts of the plants [83].

In forages, selenium content varies between 0.1 and 0.5 ppm, higher concentrations increasing the risk of animal poisoning. Plants can be classified as *i.* seleniferous (with a high content of selenium), *ii.* selenium accumulating plants and *iii.* plants with a medium selenium content. Seleniferous plants can accumulate up to 20.000 ppm of the metalloid and are represented by *Astragalus* species. Plants that display a medium content of selenium, such as *Aster, Gutierrezia* and *Atriplex* are considered to be toxic to animals [83].

Another classification refers to plants as hyperaccumulators, secondary-accumulators and non-accumulators [84]. Species from *Fabaceae* family are considered to be hyperaccumulators of selenium. Moreover, plants such as *Astraglus, Neptunia, Stanleya, Xylorhiza* etc. are also in this group. Plants from the secondary-accumulators group can store up to 100-1000 mg Se/kg dry weight without displaying toxic effects. In this group are included plants like *Brassica napus, Helianthus spp., Medicago sativa* etc. The non-accumulators store below 100 mg Se/kg dry weight and are represented by crops and grasses [84]. Plants that can store high selenium concentrations without toxic effects can be used for selenium decontamination of water and soils [85]. According to Mehdi et al. cereal plants deposit selenium in the seeds in the Semet form [83].

Various species from the *Brassicaceae* family can accumulate high concentration of selenium and thus can be used for selenium decontamination and also for human health by converting selenium into a bioactive element. In this regard, broccoli can decrease inorganic selenium toxic effects by changing into the organic forms Se-methylselenocysteine, Se-methylselenomethionine, and gamma-glutamyl-Se-methyl selenocysteine [86]. Another source of selenium are the Brazil nuts that encompass around 200 mg of Se in 1 g of nuts [87].

Iodine

Iodine is an essential element for humans and has a vital role in the thyroid function [88]. Iodine concentration in plants is influenced by the soil type, the fertilizer used and also by climate. According to Soetan et al., in farm animals, iodine deficiency has the highest prevalence, and in order to prevent this, iodine feed supplementation is required (0.35 mg/kg of iodine) [1]. An important source of iodine are the seaweed especially the ones near the coral reefs [89]. Madhusudan and coworkers reported that in seaweed, the highest iodine content was found in brown algae with values between 1500 and 8000 ppm for dry kelp and between 500 and 1000 ppm for dry *Fucus* [90]. In the case of green and red algae, the iodine amount is lower, ranging from 100 to 300 ppm in the dry material. It is estimated that the recommended daily iodine intake for adults is around 150 μg/day and by consuming 1 g of brown algae in dry form, a higher dose of 500-8000 μg could be provided.

In conformity with the World Health Organization (WHO) iodine content is higher in dry foods compared to the fresh one. In cereal grains, iodine content varies between 34 – 92 μg/kg, in dry fruits between 62-277 μg/kg and in dry vegetables around 204-1636 μg/kg [89].The literature is poor in data regarding the effect of iodine in plants and also the concentration of iodine in various plants is not well established.

Manganese

Manganese is a cofactor for several enzymes and has an important role in the amino acids formation [1]. Furthermore, the micronutrient is necessary for chlorophyllsynthesis. Manganese is found in the mitochondria and it is supposed to have a role in the oxidative phosphorylation, and also in mucopolysaccharides and glycoproteins synthesis due to the fact that it is a potent activator of glycosyltransferase enzymes. Manganese can be found in various nuts and seeds, whole grains, legumes and tea [1]. Several plants from Pakistan, from Haripur basin were evaluated in terms of their concentrations in different minor and major element [91]. The authors indicated that the highest concentration of manganese was 105.56 ppm in *Achyranthes aspera* L. and the lowest was 32.64 ppm in *Justicia adhatoda* In a study conducted on medicinal plants from Egypt, manganese levels were around 44.6 and 339 ppm [91]. In a recent study performed by Tarak and coworkers, the micronutrients (Cu, Fe, B, Mo, Mn and Zn) content was established in *Camellia sinensis* [77]. The plant is mostly cultivated in countries from Asia and depending on the preparation method, four major types of tea can be obtained, namely black, green, Oolong and Pu-erh tea. Manganese average content in black tea varies between 0.28 and 7934 μg/g, whereas in the green tea, the content is lower, between 0.55 to 3358.26 μg/g. In the Oolong tea, manganese content is between 101.95 and 2368μg/g and for the Pu-erh tea, 615 to 1268μg/g. Furthermore, manganese was present in different tea infusions. According to WHO, the daily average dose needed per person ranges between 3 to 9 mg and by consuming tea infusion three times a day, manganese intake amount varies between 2.13 and 7.89 mg. The authors concluded that tea infusion represents a good source of manganese [77].

Nickel

Nickel is an element reported to possess various roles, including cofactor for several enzymes, sustains the membrane structure and the metabolism of nucleic acid [1].

Jabeen et al., conducted a study on various plants from Pakistan. They established that the maximum concentration of nickel was found in *Cannabis sativa* (15.8 ppm) while the lowest accumulation was recorded for *Convolvulus arvensis* (2.6 ppm) [72]. Demirezen and Aksoy performed a study on different vegetables from Turkey, Kayseri area, in order to determine the concentration of different heavy metals such as cadmium, copper, lead, nickel and zinc [92]. Nickel concentration in vegetables was higher in the urban area compared to the rural area of Kayseri. The highest concentration of nickel was recorded for cucumber (13.45 μg/g), followed by lettuce (6.3 μg/g), pumpkin (5.6 μg/g), peppermint (5.1 μg/g), onion and eggplant (4.6 μg/g) etc. The lowest concentration of nickel was found in the rural area, in tomato (0.44 μg/g). The data obtained suggest that nickel concentration in these plants is lower than the recommended dietary intake according to WHO [92].

Molybdenum

Molybdenum is cofactor for different enzymes and has a role in nitrogen fixation. In the metabolic process of leguminous plants, molybdenum has a role in nitrate assimilation and nitrogen fixation [1]. For humans it is an essential element but in high doses it may induce negative effects. Moreover, molybdenum is considered to decrease the incidence of dental caries and to have beneficial effects on people with asthma. The WHO recommends for adults a daily dose of 0.15 – 0.5 mg of the microelement. The average content of molybdenum obtained from tea infusion (consumed three times a day) is estimated to be around 0.02 – 0.17 mg [77]. Many studies conducted on tea and tea infusion showed mainly a low content of molybdenum or the absence of it, probably due to the poor solubility [77]. The literature is scarce in data regarding molybdenum effect in plants. More studies are required in order to understand the mechanisms and the implications of this element in vegetal species.

THE ROLE OF MAIN MACRONUTRIENTS FOUND IN PLANTS WITHIN THE HUMAN BODY

Calcium is a divalent cation, an essential element of the human body composition (the 5th after oxygen, carbon, hydrogen and nitrogen). It is required for the proper functioning and health of the body. It accounts for approximately 1.9% of the total body weight, whereas 99% of the body calcium is found in the skeleton (calcium hydroxiapatite), 1% in the teeth and soft tissues and 0.1% in the extracellular fluid and plasma [93, 94]. Calcium intake, absorption and excretion depends on: vitamin D status, sodium/protein intake, age, childhood, pregnancy, lactation, menopausal status in women, elderly men and some co-administered drugs for a chronic treatment (> 3 months) such as steroids, proton pomp inhibitors, low molecular height heparins and antiepileptic drugs [93, 95].There are wide variations in calcium intake between countries due to their common diet: animal protein intake and dairy product consumption, Asia having the lowest calcium intake and North America and Europe the highest intake [93, 96].

Calcium plays several important roles throughout the human body (actions that require and depend on vitamin D status) such as: maintaining the integrity and rigidity of skeleton, essential for several metabolic processes in the body, assuring the neuromuscular and other cells' functions, mediator of hormonal effects on targeted organs, maintaining the cardiovascular health. [95, 96] People with calcium deficiency might have a greater risk of fractures, bone pathologies and also can develop osteoporosis, although not all the cases of osteoporosis can be attributed to calcium deficiency [93]. Sodium can induce a negative effect on the absorption of calcium, due to their competition for reabsorption in the renal tubules [93, 97].

Excess of calcium intake (> 3g/day for adults or long periods of treatment), hypercalcemia, can cause renal insufficiency, vascular and soft tissues calcifications, hypercalciuria with kidney stones and it might lower the absorption of iron and zinc [93].

Table 1. Daily calcium recommended doses (mg/day): [97, 98]

Authority	Infants and children	Adolescents (10 – 18 years old)	Females (19 – menopause)	Females post menopause	Males (19 – 65)	Males > 65	Pregnant women	Lactating women
U.S. National Institutes of Health	200 - 1000	1000 - 1300	1000	1200	1000	1200	1000 – 1300	1000 - 1300
EFSA European Food Safety Authority	280 – 800	800 – 1150	1000 – 950	950	1000 – 950	950	1000	1000

Magnesium is an essential dietary nutrient, an alkaline metal, present in the human body from 760 mg at birth to 25g at adulthood [93, 99]. From the total body content, 30-40% of magnesium is present in the muscles and the soft tissues, 50-60% in the skeleton and 1% in the extracellular fluid. The high deposits of magnesium in the bones are a continuous source of magnesium in the plasma in case of deficiency, percent that will decrease in elderly [93, 99]. It is a cofactor for more than 300 enzymatic reactions, especially for those involving the energy metabolism [97, 99]. It is responsible for the maintenance of the electrical potential of nervous tissues and cell membranes and is indispensable for the regulation of potassium fluxes and calcium metabolism (hypomagnesaemia can induce hypocalcaemia and hypokalaemia leading to neurological and cardiac manifestations, in severe cases). It is involved in the synthesis of proteins, DNA and RNA. [93, 95, 100]. Magnesium deficiency is the 2nd most common problem in developed countries, the 1st being the vitamin D deficiency, because of their poor diet in magnesium (grains are poor in magnesium compared with nuts and leafy vegetables) [93].

Early depletion of magnesium induces neurologic and/or neuromuscular defects resulting in anorexia, nausea, lethargy, weight loss, muscular weakness and spasms, hyperirritability, hyper excitability, tetany, and in severe cases convulsions, cardiac arrhythmia, and pulmonary edema

[93, 100]. High zinc intakes (> 140mg/day) or dietary fiber (40-50g/day, except: cellulose-rich products) can decrease magnesium absorption. Also high calcium (>2.5 g/day intakes) and sodium intake can lower the magnesium status within the body [93, 99].

Table 2. Daily magnesium recommended doses (mg/day): [97, 98]

Authority	Infants and children	Adolescents (10 – 18 years old)	Females (19 – menopause)	Females post menopause	Males (19 – 65)	Males > 65	Pregnant women	Lactating women
U.S. National Institutes of Health	30 – 130	130 - 410	300 – 320	320	420	420	350 – 400	310 - 360
EFSA European Food Safety Authority	80 – 230	300	300	300	350	350	300	300

Higher doses of magnesium, hypermagnesaemia, can induce nausea, hypotension and diarrhea [99].

Potassium is an essential nutrient (not produced naturally by the human body), the most abundant element of the intracellular fluid [101]. This electrolyte plays an important role in maintaining the cell function, especially in excitable cells such as: muscles and nerves. As well as magnesium, it is widely distributed in foods (fruits and vegetables, although cereals and meat have a lower concentration in potassium). During the last decades, the Western alimentation practices (lower consumption of vegetables and fruits; higher consumption of cereals and processed food) lead to a diet higher in sodium and lower in potassium [93, 101].

The body fluids of an adult account for 50 – 70% of his body weight (42 L) and are composed of water and electrolytes (potassium, sodium, chloride and phosphorus). Potassium and phosphorus are mostly present in the intracellular fluid, whereas sodium and chloride are dominant in the extracellular fluid. Approximately 98% of the total gastrointestinal absorbed potassium is stored within the cells (140 mEq/L) and 2% being in the extracellular fluid (5 mEq/L).The normal plasma levels for potassium in adults are 3.5 – 5.0 mEq/L (for an average dietary intake of 80 – 200 mEq/day). The potassium homeostasis depends on the dietary intake (absorption versus excretion) and on the internal balance (distribution between the intracellular space and the extracellular space). This homeostasis is maintained primarily by the kidneys (renal mechanisms). Potassium is stored mostly in the muscles and lower amounts are present in the bones, skin and red cells [93, 97, 101].

The roles of potassium within the body are wide and numerous. Most of the body potassium concentration is found in the intracellular space of the skeletal muscle, having a key role in cell functioning, inducing the cellular membrane potential and regulating the electrolyte and fluid balance. It is responsible for the nerve impulse transmission, skeletal and cardiac muscle contractions, heart function, blood pressure (the increased prevalence of hypertension being related to a lower potassium intake rather than to a higher sodium intake). It induces vasorelaxation in the blood vessels, regulates the intestinal motility, glucose and insulin metabolism [102]. Hypokalemia has been found to induce impaired glucose tolerance (due to reduced insulin secretion) and also impaired glucose transportation at cell membranes and thus, a higher risk of diabetes (interference with the functioning of beta cells) [103]. Moreover, it is involved in hormone secretion, electrical activity of the heart, mineral-corticoid influence, enzyme activation metabolism of carbohydrates and proteins, growth and muscles building [93].

In case of potassium deficiency, hypokalemia (< 3.5mEq/L in the serum), symptoms are: fatigue, muscle weakness and spasms, slow reflexes, nausea, vomiting, dry skin, cardiac arrest, impaired glucose metabolism, hyperglycemia, bone fragility, insomnia, decreased urine production, cardiac arrhythmias [97]. Excess of potassium in the blood, hyperkalemia (> 5 mEq/L), a situation less common than hypokalemia, can induce: respiratory insufficiency, weakness and even paralysis, ventricular tachycardia and fibrillation [97]. Except diet, several drugs can impact the potassium levels, such as: angiotensin-converting enzyme inhibitors, angiotensin receptor blockers, some diuretics (spironolactone, amilorid, eplerenone) can induce hyperkalemia while others (furosemide, indapamide, hydrochlorothiazide) can induce hypokalemia [103].

Table 3. Daily potassium recommended doses (g/day): [97, 98]

Authority	**Infants and children**	**Adolescents (10 – 18 years old)**	**Females (19 – menopause)**	**Females post menopause**	**Males (19 – 65)**	**Males > 65**	**Pregnant women**	**Lactating women**
U.S. National Institutes of Health	0.4 – 3.8	3.8 – 4.7	4.7	4.7	4.7	4.7	4.7	5.1
EFSA European Food Safety Authority	0.75 – 1.8	1.8 – 3.5	3.5	3.5	3.5	3.5	3.5	4.0

Nitrogen is the most plentiful element in the atmosphere. Its molecular form, the inert state of nitrogen (N_2), cannot be used by the vast majority of the living organisms so it has to be transformed into reactive nitrogen, in order to be used by the living organisms. It is essential for growth, although in high concentrations it can induce several negative effects on the environment (risks to human health, with economic and social consequences). In the industrialized countries nitrogen is present in very high amounts (in the soil, water and air) inducing environmental, economic and health problems, compared with in developing countries, were the nitrogen can be found in small amounts, with negative consequences on agriculture (less food production to sustain the population), land degradation, soil erosion. So, whether in excess or in deficiency, nitrogen concentration needs monitoring and further research. It is essential for plants animals and humans to survive. The human beings cannot breathe nitrogen. They are absorbing nitrogen through the alimentation. The human body eliminates it by converting ammonia to urea, which will be excreted in the urine (the way through which nitrogen returns back to the environment) [104, 105]. It has an important role in: amino acid synthesis that will later generate proteins, nucleic acid production which will later generate DNA and RNA, hormone synthesis, brain functioning, immune system, cell replacement and tissue repair [93, 97, 104].

Phosphorus is, after calcium, the second most highly present essential mineral in the human body. It is synthesized by several biochemical pathways: glycolysis and beta oxidation. From the total amount in the body, 80 – 90% is present in the bones and in the teeth (as hydroxyapatite), the rest being found in the extracellular fluid, soft tissues, erythrocytes. The serum and plasma contain very small fractions of phosphorus and changes in the serum phosphate levels cannot reflect entirely the body's total amount of phosphorus. The phosphate homeostasis is maintained by the action of PTH (parathyroid hormone), vitamin D_3 and phosphatonins [106, 107]. It plays an important role in biologic processes (such as energy metabolism and bone mineralization) and is a structural framework for DNA and RNA. Phosphorus is a component of cell membranes and walls and it is commonly found in the body as phosphate. It is involved in protein synthesis, maintenance of blood pH and acid – base balance. Several hormones and enzymes contain phosphorus as a structural component and hemoglobin depends on it in order to proper functioning [93, 97, 107].

A normal diet provides 20 mg/kg/day of phosphorus. Deficiencies in phosphorus (< 2.5mg/dl serum phosphate) were observed in cases of decreased or inadequate dietary intake, high urinary excretion by the kidneys, tumor-induced osteomalacia and rickets. Deficiencies are rare (widely present in food) and can affect normal bone and teeth development, in children [97]. Hyperphosphatemia (>5mg/dl) was observed in patients with alteration of renal function, defective phosphatonin, increase dietary intake and can induce renal problems. Patients with chronic kidney disease develop hyperphosphatemia due to decreased elimination, increase of calcitriol (vitamin D) levels and secondary hyperparathyroidism [97].

Table 4. Daily phosphorus recommended doses (mg/day): [97, 98]

Authority	Infants and children	Adolescents (10 – 18 years old)	Females (19 – menopause)	Females post menopause	Males (19 – 65)	Males > 65	Pregnant women	Lactating women
U.S. National Institutes of Health	100 – 500	500 – 1250	1250 – 700	700	1250 – 700	700	700	700
EFSA European Food Safety Authority	160 – 440	440 – 640	640 – 550	550	640 – 550	550	550	550

Sulphur's primary sources in human being are SAAs (sulphur containing amino acids) obtained from diet. Metionine and cystine are part of SAAs, substances essential for humans, from which metinone cannot be synthesized in the organism. More than 70% of the amount of sulphur in plants is present as amino acids. The sulphur containing aminoacids are: methionine, cysteine, homocysteine, taurine, thiamin, biotin, ALA (alpha-lipoic acid), coenzyme A, glutathione, glucosamine/chondrointin sulfate, fibrinogen, heparin etc. All these SAAs are synthesized from methionine (except: thiamin and biotin). MSM (methyl sulfonylmethane) is another source of sulphur that is taken from the diet [108, 109]. Sulphur can be successfully used in several pathologies such as: dermatological disorders (acne, dandruff, skin rashes, dermatitis, psoriasis), wound healing, acute exposure to radioactive material (antidote), imunodepression, osteoarthritis, fatty liver, atherosclerosis, bone demineralization, degenerative disorders. Also, it is very important for athletes in order to reduce the pro-inflammatory status caused by their training [93, 97, 108].

THE ROLE OF MAIN MICRONUTRIENTS FOUND IN PLANTS WITHIN THE HUMAN BODY

Chlorine is involved in the regulation of osmotic balance and water/salt metabolism. The ionic form is chloride and it is necessary for human body in large quantities. It is present in high amounts in the skin (30-60%), but also in the intercellular fluid, blood and bones. It is derived exclusively from diet and it is the most abundant anion in humans. It is often added to water in order to clean it, as it has the potential to inhibit the spreading and transmission of several infections (hepatitis, typhoid etc) and it forms the hypochlorite ($HOCl^-$), a potent antiseptic agent with broad spectrum activity [110]. It is vital for several physiologic functions in the human body, such as: fluid retention and secretion, osmotic balance, pH balance, immunity, optimal activity of phagosomes, production of gastric juice (hydrochloric acid), healthy functioning of brain, nerves and muscles, synthesis of blood plasma and helps the body to eliminate toxins, carbon dioxide and liver fat [110].

Table 5. Daily chlorine recommended doses (g/day): [97, 98]

Authority	**Infants and children**	**Adolescents (10 – 18 years old)**	**Females (19 – menopause)**	**Females post menopause**	**Males (19 – 65)**	**Males > 65**	**Pregnant women**	**Lactating women**
U.S. National Institutes of Health	0.18 – 1.9	1.9 – 2.3	1.9 – 2.3	1.8	1.9 – 2.3	1.8	2.3	2.3
EFSA European Food Safety Authority	On-going	On-going	On-going	On-going	On-going	On-going	On-going	On-going

Excess of chlorine in the human body can be harmful as it promotes water retention in tissues and organs, edema, elevation of blood pressure, dry cough, dyspepsia with negative consequences. The lack of chlorine can cause drowsiness, muscle weakness, dry mouth, loss of appetite, memory, hair or teeth [110].

Iron is present in the human body in: erythrocytes (hemoglobin), muscles (myoglobin), iron-containing enzymes and cytochromes. It is reversibly stored in the liver as ferritin and haemosiderin and is transported through the body by transferrin. It is not actively excreted from the body in urine or intestines and it is lost only with cell death (apoptosis) of the skin, intestines, urinary tract and airways. It has several vital functions such as: carrier of oxygen to the tissues (hemoglobin) from the lungs, transporter for electrons within the cells, component of enzyme systems in various tissues, transport of energy within the cell and to mitochondria, synthesis of steroid hormones and bile acids, detoxification of foreign substances within the liver, signal controlling of some neurotransmitters (dopamine and serotonin) in the brain [97, 111, 112].

There are two kinds of dietary iron: haem iron (found in meat, poultry, fish) and non-haem iron (found in cereals, vegetables, legumes, fruits). The non-haem iron is the main dietary form of iron. Substances that can inhibit the dietary absorption of iron are: calcium (dairy products), bran products, oats, rice, cocoa, soya, beans, tea, coffee, some spices, red wine. Iron dietary absorption can be increased by ascorbic acid (some fruit juices, fruits, potatoes, vegetables), meat, fish, sea food [93].Population at risk for iron deficiency is represented by: infants, children, adolescents, pregnant and lactating women. A diagnosis of iron deficiency must include a search for some pathological cases of blood loss (tumors: gastro-intestinal tract or uterus; achlorhydria due to intensive gastric surgery or gluten enteropathy). Iron deficiency anemia is the most common nutritional disorder [93, 111].

The prevention of iron deficiency in young children is highly important and urgent, both in developing countries and in developed ones, because of the risk of impaired brain development even in mild forms of iron deficiency. Moreover, supplementation of the diet with iron at a later date, after the symptoms has been installed (functional brain defects) cannot be reversed [97, 111].

Table 6. Daily iron recommended doses (mg/day) by the: [97, 98]

Authority	Infants and children	Adolescents (10 – 18 years old)	Females (19 – menopause)	Females post menopause	Males (19 – 65)	Males > 65	Pregnant women	Lactating women
U.S. National Institutes of Health	0.27 – 10	11 – 8	8 – 18	8	8	8	27	9 - 10
EFSA European Food Safety Authority	7 – 11	11 - 13	13 – 16	11	11	11	16	16

Zinc is present in all the body tissues and fluids. The skeletal muscles and the bone mass account for approximately 60% of the total body amount.

Table 7. Daily zinc recommended doses (mg/day): [97, 98]

Authority	Infants and children	Adolescents (10 – 18 years old)	Females (19 – menopause)	Females post menopause	Males (19 – 65)	Males > 65	Pregnant women	Lactating women
U.S. National Institutes of Health	2 – 5	8 – 11	8	8	11	11	11 – 12	12 - 13
EFSA European Food Safety Authority	2.9 – 7.4	7.4 – 14.2	11.9 – 12.7	11.9 – 12.7	9.4 – 16.3	9.4 – 16.3	9 – 14	10.5 - 16

There are no zinc deposits in the body in the conventional sense and zinc plasma concentrations can be maintained at a normal level over several months. High losses of zinc are due to: starvation, muscle catabolism (physical exercises), perspiration (high temperatures), food intake, fever, infection and pregnancy [93, 97, 113]. By reducing the phytate content and by adding animal protein to the diet, the absorption of zinc from the diet can be increased [93]. It has a central role in growth, cell division, protein synthesis, immunity, height and weight increase, catalytic role for several enzymes within the human body, signal transduction. Changes in immune response and serum lipid pattern were observed after zinc supplementation and high zinc intake (> 50mg/day) can affect copper status [114].

Selenium can be found in the human body in the following percentages: 30% in the liver, 30% in the muscles, 15% in the kidneys and around 10% in the blood plasma. High amounts of the tissue selenium can be found in proteins in the body. The main roles of selenium in the human body are: body tissues' protection against oxidative stress, protection against infections, skeletal and cardiac muscle metabolism, growth and development.

Table 8. Daily selenium recommended doses (mcg/day): [97, 98]

Authority	Infants and children	Adolescents (10 – 18 years old)	Females (19 – menopause)	Females post menopause	Males (19 – 65)	Males > 65	Pregnant women	Lactating women
U.S. National Institutes of Health	15 – 30	30 – 55	55	55	55	55	60	70
EFSA European Food Safety Authority	15 – 35	35 - 70	70	70	70	70	70	85

The two different families of selenium containing enzymes: glutathione peroxidase and thioredoxinreductase are involved in the controlling of reactive oxygen metabolites production in the tissues. They are essential at low doses for assuring the innate immune response but are highly toxic in excess. Associated with vitamin A, selenium protects the cell membranes against oxidative stress. Low selenium status can have a modifying influence on thyroid hormone balance [115, 83]. There is no clear evidence on selenium depletion, the clinical manifestations being not so well defined and include muscular weakness, myalgia and higher risk of developing viral infections. Excess of selenium in the body (> 900 mcg/day) can induce hair loss, gastrointestinal disorders, structural changes in the keratin of nails and hair [116].

Iodine has a single, well-known, physiological role in the human body: it is involved in the synthesis of thyroid hormones by the thyroid gland. The dietary intake of iodine is determined by the thyroid gland, for a normal thyroxine (T_4) production. All the biological actions of iodide are attributed to thyroid hormones. The most important indexes for iodine intake are: serum T_4, TSH levels and urinary iodine. The major implications of thyroid hormones in the human body are in the: growth and development of brain and central nervous system and control of several metabolic processes (e.g.: carbohydrates, fat, protein, minerals, vitamins). Iodine deficiency is obligatory to be diagnosed and treated in: women at reproductive age, pregnant and lactating women, children under 3 years, because it can lead to irreversible brain and central nervous system damage and lesions and thus, to mental retardation (cretinism) [93, 117].

Table 9. Daily iodine recommended doses (mcg/day): [97, 98]

Authority	**Infants and children**	**Adolescents (10 – 18 years old)**	**Females (19 – menopause)**	**Females post menopause**	**Males (19 – 65)**	**Males > 65**	**Pregnant women**	**Lactating women**
U.S. National Institutes of Health	110 – 90	120 - 150	150	150	150	150	220	290
EFSA European Food Safety Authority	70 – 90	90 – 130	130 – 150	150	130 – 150	150	200	200

Excess of iodine can be harmful, inhibiting the secretion and release of thyroid hormones, with dramatic consequences for the proper functioning of the human body [93].

Boron is present in nature as borate substances with other elements. It is widely spread in the nature, having the following functions and roles in the human body: growth, bone density and development, embryonic development, central nervous system development, wound healing, hepatic metabolism, innate immunity, synthesis of testosterone and estrogen, anti-inflammatory, antithrombotic, antioxidant, anticancer, antimicrobial agent, hypocholesterolemiant, hypoglucidic. Serum calcium and magnesium can improve with boron supplementation. Low boron profile can induce: osteoporosis, cognitive deterioration, immunodepression and can exacerbate arthritis. The optimal intake of boron seems to be unknown and vary between 3 – 10 mg/day in adults. Toxicity of boron (> 20mg/day in adults) can induce congestion, inflammation, renal toxicity, dermatitis, edema etc. [118, 119].

Copper is a vital metal ion, required in very small amounts for the human well-being. It is the 3rd most abundant trace metal after iron and zinc. It is present in every tissue of the body, but it is stored primarily in the liver and then in the brain, heart, muscles and kidneys. Among its biological roles the most important are: cofactor for the enzymes that catalyze redox reactions, maintaining the strength of skin, blood vessels, connective tissue, interferes with the synthesis of hemoglobin, melanin, myelin and thyroid hormones, innate immunity, essential substance for gene expression (it activates and suppress gene transcription). It is directly and indirectly involved in the pathogenesis of several neurological diseases, such as: Alzheimer, Huntington, Menken, Parkinson, Wilson and amyotrophic lateral sclerosis [120]. And it can either act as an antioxidant or pro-oxidant, in function of its concentration, so it is highly important to maintain the dietary balance of copper, together with zinc and manganese. Deficiency of copper can induce: anemia, neutropenia, bone abnormalities, fractures, high risk of infections, impaired weight gain, increased of total serum cholesterol and LDL cholesterol, decrease of glucose tolerance, neurological problems, poor thyroid function and joint problems [121]. In the latest years, the concerns of the nutritionists were not related to its deficiency, but rather to its toxicity due to the high use of copper water pipes and copper cookware. High copper intake can cause: nausea, vomiting, cramps, headache, diarrhea, metallic taste and can induce liver and kidney problems [120, 123].

The recommended doses vary between 340 mcg (children) to 900 mcg (in adults) in USA. It is important to mention that the standard diet is sufficient to supply the adequate daily needs for all ages [93].

Table 10. Daily copper recommended doses (mcg/day): [97, 98]

Authority	Infants and children	Adolescents (10 – 18 years old)	Females (19 – menopause)	Females post menopause	Males (19 – 65)	Males > 65	Pregnant women	Lactating women
U.S. National Institutes of Health	200 – 440	700 – 900	900	900	900	900	1000	1300
EFSA European Food Safety Authority	400 – 1000	1300 – 1600	1300	1300	1600	1600	1500	1500

Manganese is an essential nutrient for intracellular activity in humans, being co-factor for several enzymes (arginase, glutamine synthase) and metalloproteins (pyruvate carboxulase and manganese superoxide dismutase). It is essential for human development, energy metabolism, immunological system, central nervous system function, blood clotting, antioxidant processes, bone and connective tissue growth, neuro-transmitters' synthesis and metabolism, antioxidant processes, reproductive hormone function [123, 97].Manganese deficiency is rare, compared with its toxicity (poisoning). Deficiency in manganese is associated with growth impairment, bone formation, reduced fertility, birth defects, dermatitis, alteration of lipid, protein and carbohydrate metabolism [97, 123, 124].

Manganese excess in the human body inducesneurotoxicity ("manganism") a psychiatric and motor disturbances pathology. Except for the brain toxicity, it can accumulate in the liver, kidneys, pancreas and bones causing hepatic cirrhosis, polycythemia and symptoms of Parkinson's disease [123 - 125].

Table 11. Daily manganese recommended doses (mg/day): [97, 98]

Authority	Infants and children	Adolescents (10 – 18 years old)	Females (19 – menopause)	Females post menopause	Males (19 – 65)	Males > 65	Pregnant women	Lactating women
U.S. National Institutes of Health	0.003 – 1.9	1.9 – 2.2	1.6 – 1.8	1.8	2.3	2.3	2.0	2.6
EFSA European Food Safety Authority	0.02 - 1.5	1.5 – 3.0	1.5 – 3.0	3.0	3.0	3.0	3.0	3.0

Nickel has a fundamental role for humans, although is a "double-faced" element for life. There were no nickel containing enzymes reported in humans. It catalysis biochemical reactions in the active site of metalloenzymes (urease, hydrogenase, co-dehydrogenase etc.). It is essential for the colonization and growth of many parasites that infect the human body, being used by the gut microbiota. Nickel deficiency is extremely rare, accumulation and toxic effects of nickel being more often reported. At the cellular level the toxic effects are represented by changes in metabolism, cell proliferation, oxidative stress, inflammation and apoptosis. Accumulation of nickel in the human body can induce allergic/systemic contact dermatitis, carcinogenesis (lung and nasal cancers), infection diseases and development of intestinal microbes [126, 127].Nickel requirements for humans can vary between 5 - 50mcg/day and it is usually assured by the diet, as nickel deficiency has never been reported [127].

Molybdenum is a life-sustaining element, needed by plants, animals and humans in low concentrations. It has an important role in food digestion (synthesis of enzymes), energy production, elimination of toxic compounds from the body. There are three enzymes in the human body that require molybdenum: xanthine oxidase (that regulates uric acid synthesis), aldehyde oxidase (which regulates the synthesis of steroid hormones, vitamin A metabolism and is important for the lung and blood circulation) and sulfite oxidase (which helps the body eliminate the sulfite, a neurotoxic substance) [128]. The substance tetrathiomolybdate is used to reduce copper absorption and to increase copper excess elimination from the body. Moreover, molybdenum influences the metabolism of other elements as well, such as: phosphorus, sulfur, potassium, iron, zinc and iodine. It showed promising results for the treatment of several cancers (metastatic kidney cancer, colorectal cancer and breast cancer) [129]. Lack of molybdenum can induce nitrogen deficiency and impaired growth, kidney stone and renal failure. It is harmful in high concentrations and its excess can cause gout symptoms, anorexia, fatigue, headache, acute psychosis, major depression, learning problems and post-traumatic stress disorder [129].

Table 12. Daily molybdenum recommended doses (mg/day): [97, 98]

Authority	**Infants and children**	**Adolescents (10 – 18 years old)**	**Females (19 – menopause)**	**Females post menopause**	**Males (19 – 65)**	**Males > 65**	**Pregnant women**	**Lactating women**
U.S. National Institutes of Health	2 – 22	34 – 43	45	45	45	45	50	50
EFSA European Food Safety Authority	10 – 30	30 – 65	65	65	65	65	65	65

REFERENCES

[1] Soetan, K. O., Olaiya, C. O., & Oyewole, O. E. (2010). The importance of mineral elements for humans, domestic animals and plants. *African Journal of Food Science*, *4*(5), 200–222. https://doi.org/10.1186/s12302-017-0116-y.

[2] White, P.J., Brown, P.H. (2010). Plant nutrition for sustainable development and global health. *Ann Bot*, 105(7): 1073–1080, doi: 10.1093/aob/mcq085.

[3] Typical concentrations sufficient for plant growth. After E. Epstein. (1965). "Mineral metabolism" pp. 438-466. in: *Plant Biochemistry* (J. Bonner and J.E. Varner, eds.) Academic Press, London.

[4] Merchant, S.S. (2010). The Elements of Plant Micronutrients. *Plant Physiol*, 154(2): 512–515, doi: 10.1104/pp.110.161810.

[5] Cetinkaya, H., Koc, M., Kulak, M. (2016). Monitoring of mineral and polyphenol content in olive leaves under drought conditions: Application chemometric techniques. *Industrial Crops and Products*, 88:78-84.

[6] Khan, M., Iqbal, R., Iqbal, N., Masood, A., Mobin, M., Anjum, N.A., and Khan, N.A. (2016). Modulation and significance of nitrogen and sulfur metabolism in cadmium challenged plants. *Plant Growth Regulation* 78(1): 1-11.

[7] Wink, M. (2013). Evolution of secondary metabolites in legumes (Fabaceae). *South African Journal of Botany*, 89: 164-175.

[8] Maróti, G., Kondorosi, E. (2014). Nitrogen-fixing *Rhizobium*-legume symbiosis: are polyploidy and host peptide-governed symbiont differentiation general principles of endosymbiosis? *Frontiers in Microbiology*, 5:326.

[9] Oldroyd, G. E., Murray, J.D., Poole, P.S., Downie, J.A. (2011). The rules of engagement in the legume-rhizobial symbiosis. *Annual Review of Genetics*, 45:119-144.

[10] Bouchenak, M., Lamri-Senhadji, M. (2013). Nutritional quality of legumes, and their role in cardiometabolic risk prevention: a review. *Journal of Medicinal Food*, 16(3):185-198.

[11] Vega-Gálvez, A., Miranda, M., Vergara, J., Uribe, E., Puente, L., Martínez, E.A. (2010). Nutrition facts and functional potential of quinoa (*Chenopodium quinoa* Willd.), an ancient Andean grain: a review. *Journal of the Science of Food and Agriculture*, 90(15): 2541-2547.

[12] Valdivia-López, M.Á., Tecante, A. (2015). Chia (*Salvia hispanica*): a review of native Mexican seed and its nutritional and functional properties. *Advances in Food and Nutrition Research*, 75: 53-75.

[13] Montoya-Rodríguez, A., Gómez-Favela, M.A., Reyes-Moreno, C., Milán-Carrillo, J., and Elvira González de Mejía, E.G. (2015). Identification of bioactive peptide sequences from amaranth (*Amaranthus hypochondriacus*) seed proteins and their potential role in the prevention of chronic diseases. *Comprehensive Reviews in Food Science and Food Safety*, 14(2):139-158.

[14] López, D.N., Galante, M., Raimundo, G., Spelzini, D., Boeris, V. (2019). Functional properties of amaranth, quinoa and chia proteins and the biological activities of their hydrolyzates. *Food Research International*, 116:419-429.

[15] Chrysargyris, A., Panayiotou, C., Tzortzakis, N. (2016). Nitrogen and phosphorus levels affected plant growth, essential oil composition and antioxidant status of lavender plant (*Lavandula angustifolia* Mill.). *Industrial Crops and Products*, 83:577-586.

[16] Balemi, T., Negisho, K. (2012). Management of soil phosphorus and plant adaptation mechanisms to phosphorus stress for sustainable crop production: a review. *Journal of Soil Science and Plant Nutrition*, 12(3):547-562.

[17] Konieczyński, P., Wesołowski, M. (2007). Determination of Zinc, Iron, Nitrogen and Phosphorus in Several Botanical Species of Medicinal Plants. *Polish Journal of Environmental Studies*, 16(5):785-790.

[18] Noori, N., Sims, J.J., Kopple, J.D., Shah, A., Colman, S., Shinaberger, C.S., Bross, R., Mehrotra, R., Kovesdy, C.P., Kalantar-Zadeh, K. (2010). Organic and inorganic dietary phosphorus and its

management in chronic kidney disease. *Iranian Journal of Kidney Diseases*, 4(2):89-100.

[19] Kumar, V., Sinha, A.K., Makkar, H.P.S., Becker, K. (2010). Dietary roles of phytate and phytase in human nutrition: A review. *Food Chemistry*, 120(4): 945-959.

[20] Marschner, P. (2012). *Marschner's Mineral Nutrition of Higher Plants*, 3rd ed. Academic Press: London, UK, pp. 178–189.

[21] Anke, M. (2004). Potassium, In: *Elements and their Compounds in the Environment*, 2nd edition (Merian E., Anke M., Ihnat M., Stoeppler M. eds.). Wiley VCH Verlag, Weinheim,pp. 521-547.

[22] Wang M, Zheng Q, Shen Q, Guo S. The Critical Role of Potassium in Plant Stress Response. *Int. J. Mol. Sci.* 2013; 14: 7370-7390; doi:10.3390/ijms14047370.

[23] Anke, M., Bergmann, K., Lösch, E., Müller, R. (2003). Potassium intake, balance and requirement of adults, In: *Vitamins and additives in the nutrition of umans and animals* (Schubert R., Flachowsky G., Jahreis G., Bitsch R. eds.). Bundesforschungsanstalt für Landwirtschaft, Braunschweig, Germany.

[24] Antal, D.S. (2005). *Researches regarding the content in mineral elements of the medicinal plants from the Aninei Mountains*. PhD Thesis. University of Medicine and Pharmacy Târgu-Mureş (Romania).

[25] Antal, D.S., Dehelean, C., Peev, C., Anke, M. (2006). The potassium-sodium ratio of medicinal species. *Proceedings of the 4th Conference of Medicinal and Aromatic Plants of South-East European Countries*, Iaşi (Romania), 330-334.

[26] Schilcher, H., Vahlensieck, W. (2001). *Phytotherapy in Urology.* Hippokrates Verlag, Stuttgart, pp. 3-10.

[27] White, P.J., Broadley, M.R. (2003). Calcium in plants. *Annals of Botany*, 92:487-511.

[28] Yang, J., Punshon, T., Guerinot, M.L., Hirschi, K.D. (2012). Plant Calcium Content: Ready to Remodel. *Nutrients*, 4:1120-1136.

[29] Nakata, P. A. (2012). Plant calcium oxalate crystal formation, function, and its impact on human health. *Front. Biol.*, 7(3): 254–266.

[30] Weaver, C.M., Heaney, R.P., Nickel, K.P., Packard, P.I. (1997). Calcium Bioavailability from high oxalate vegetables: Chinese vegetables, sweet potatoes and rhubarb. *J Food Science*, 62(3): 524-525.

[31] Peterlik, M., Stoeppler, M. (2004). Calcium, In: *Elements and their Compounds in the Environment*, 2nd edition (Merian E., Anke M., Ihnat M., Stoeppler M. eds.). Wiley VCH Verlag, Weinheim, pp. 599-613.

[32] Fleming, K.H., Heimbach, J.T. (1994). Consumption of calcium in the U.S.: Food sources and intake levels. *J. Nutr.*, 124:1426S–1430S.

[33] Anke, M., Krämer-Beselia, K., Lösch, E., Müller, R., Müller, M., Seifert, M. (2002). Calcium supply, intake, balance and requirement of man. First information: calcium content of plant food. In: *Macro- and Trace Elements* (Anke M., Müller R., Schafer U., Stoeppler M., eds.). Schubert-Verlag, Leipzig, pp. 1386-1391.

[34] *WHO Trace Elements in Human Nutrition and Health*. (1996). World Health Organization, Geneva.

[35] Haidu, D., Negrea, A., Ianăşi, C., Antal, D., Sfirloaga, P., Kurunczi, L. (2017). Contradictory Aspects of bioaccumulation. ICP-MS, an approachable method for elemental characterization of crop medicinal plants. *Digest Journal of Nanomaterials and Biostructures*, 12(2): 391-400.

[36] Shaul, O. (2002). Magnesium transport and function in plants: the tip of the iceberg. *BioMetals*, 15: 309-323.

[37] Cakmak, I., Kirkby, E.A. (2008).Role of magnesium in carbon partitioning and alleviating photo oxidative damage. *Physiologia Plantarum*, 133: 692-704.

[38] Nielsen, F.H. (2015). Importance of plant sources of magnesium for human health. *Crop & Pasture Science*, 66:1259–1264.

[39] Rosanoff, A. (2013). Changing crop magnesium concentrations: impact on human health. *Plant and Soil*, 368:139–153.

[40] Volpe, S.L. (2013). Magnesium in disease prevention and overall health. *Advances in Nutrition*, 4:378S–383S.

[41] Anke, M., Glei, M., Groppel, B. (2000). Magnesium in the food chain of farm animals and humans. *Rekasan-Journal*, 13:13-19.

[42] Vormann, J. (2004) Magnesium. In: *Elements and their Compounds in the Environment*, 2nd edition (Merian E., Anke M., Ihnat M., Stoeppler M. eds.). Wiley VCH Verlag, Weinheim, pp. 587-598.

[43] Prkić, A., Politeo, N., Giljanović, J., Sokol, V., Bošković, P., Brkljača, M., Stipišić, A. (2018). Survey of content of cadmium, calcium, chromium, copper, iron, lead, magnesium, manganese, mercury, sodium and zinc in chamomile and green tea leaves by electrothermal or flame atomizer atomic absorption spectrometry. *Open Chem.*, 16: 228–237.

[44] Haidu, D., Párkányi, D., Moldovan, R.I., Savii, C., Pinzaru, I., Dehelean, C., Kurunczi, L. (2017). Elemental Characterization of Romanian Crop Medicinal Plants by Neutron Activation Analysis. *Journal of Analytical Methods in Chemistry*, 9748413. doi: 10.1155/2017/9748413.

[45] Chizzola, R., Michitsch, H., Mitteregger, U.S. (2008). Extractability of selected mineral and trace elements in infusions of chamomile. *Int J Food Sci Nutr*, 59(6):451–456.

[46] Prasad, R., and Shivay, Y.S. (2018). Sulphur in soil, plant and human nutrition. *Proceedings of the National Academy of Sciences, India Section B: Biological Sciences*, 88(2): 429-434.

[47] Capaldi, Flávia R., Gratão, P.L., Reis, A.R., Lima, L.W., and Azevedo, R.A. (2015). Sulfur metabolism and stress defense responses in plants. *Tropical Plant Biology*, 8(3-4):60-73.

[48] Anjum, N.A., Gill, S.S., Umar, S., Ahmad, I., Duarte, A.C., Pereira, E. (2012). Improving growth and productivity of *Oleiferous Brassicas* under changing environment: significance of nitrogen and sulphur nutrition, and underlying mechanisms. *The Scientific World Journal*, 2012:657808. doi: 10.1100/2012/657808.

[49] Mazumder, A., Dwivedi, A., du Plessis, J. (2016). Sinigrin and its therapeutic benefits. *Molecules*, 21(4):416. doi: 10.3390/molecules 21040416.

[50] Sobolewska, D., Podolak, I., Makowska-Wąs, J. (2015). *Allium ursinum*: botanical, phytochemical and pharmacological overview. *Phytochemistry Reviews*, 14(1):81-97.

[51] Borlinghaus, J., Albrecht, F., Gruhlke, F. Ch., Nwachukwu, I.D., Slusarenko, A.J. (2014). Allicin: chemistry and biological properties. *Molecules*, 19(8):12591-12618.

[52] Doleman, J. F., Grisar, K., van Liedekerke, L., Saha, S., Roe, M., Tapp, H.S., Mithen, R.F. (2017). The contribution of alliaceous and cruciferous vegetables to dietary sulphur intake. *Food Chemistry*, 234: 38-45.

[53] Raven, J. A. (2017). Chloride: Essential micronutrient and multifunctional beneficial ion. *Journal of Experimental Botany*, 68(3):359–367. https://doi.org/10.1093/jxb/erw421.

[54] Chen, W., He, Z. L., Yang, X. E., Mishra, S., Stoffella, P. J. (2010). Chlorine nutrition of higher plants: Progress and perspectives. *Journal of Plant Nutrition*, *33*(7):943–952. https://doi.org/10.1080/01904160903242417.

[55] Soetan, K. O., Olaiya, C. O., Oyewole, O. E. (2010). The importance of mineral elements for humans, domestic animals and plants. *African Journal of Food Science*, 4(5):200–222. https://doi.org/10.1186/s12302-017-0116-y.

[56] Gupta, U. C., Gupta, S. C. (2014). Sources and Deficiency Diseases of Mineral Nutrients in Human Health and Nutrition: A Review. *Pedosphere*, 24(1):13–38. https://doi.org/10.1016/S1002-0160(13)60077-6.

[57] Kobayashi, T., Nozoye, T., Nishizawa, N. K. (2018). Iron transport and its regulation in plants. *Free Radical Biology and Medicine*, pii: S0891-5849(18)32257-3. https://doi.org/10.1016/j.freeradbiomed.2018.10.439.

[58] Stein, A. J. (2010). Global impacts of human mineral malnutrition. *Plant and Soil*, 335(1): 133–154. https://doi.org/10.1007/s11104-009-0228-2.

[59] Connorton, J. M., Balk, J., Rodríguez-Celma, J. (2017). Iron homeostasis in plants-a brief overview. *Metallomics*, 9(7):813–823. https://doi.org/10.1039/c7mt00136c.

[60] Prentice, A. M., Mendoza, Y. A., Pereira, D., Cerami, C., Wegmuller, R., Constable, A., Spieldenner, J. (2017). Dietary strategies for improving iron status: Balancing safety and efficacy. *Nutrition Reviews*, 75(1):49–60. https://doi.org/10.1093/nutrit/nuw 055.

[61] Posen, J. S. (2013). Iron and vegetarian diets. *The Medical Journal of Australia*, 199(4): S11–S16. https://doi.org/10.5694/mjao11.11494.

[62] Singh, G., Kawatra, A., & Sehgal, S. (2001). Nutritional composition of selected green leafy vegetables, herbs and carrots. *Plant Foods for Human Nutrition*, 56(4):359–364. https://doi.org/10.1023/A:101187 3119620.

[63] García-Casal, M. N., Pereira, A. C., Leets, I. (2007). High iron content and bioavailability in humans from four species of marine algae. *The Journal of Nutrition*, 137(12), 2691–2695. https://doi. org/10.1093/jn/137.12.2691.

[64] Ancuceanu, R., Dinu, M., Hovaneţ, M. V., Anghel, A. I., Popescu, C. V., & Negreş, S. (2015). A survey of plant iron content—a semi-systematic review. *Nutrients*, 7(12):10320–10351. https://doi.org/10. 3390/nu7125535.

[65] T. M. Ansari, N. Ikram, M. Najam-ul-Haq, I. Fayyaz, Q. Fayyaz, I. G. and N. K. (2004). Essential trace metal (Zinc, Manganese, Copper and Iron) levels in plants of medicinal importance. *Journal of Biological Sciences*, 4(2):95–99.

[66] Guerrero, J., Martınez, J., & Isasa, M. (1998). Mineral nutrient composition of edible wild plants. *Journal of Food Composition and Analysis*, 11(4): 322–328.

[67] Hafeez, B. (2013). Role of Zinc in Plant Nutrition- A Review. *American Journal of Experimental Agriculture*, 3(2): 374–391. https://doi.org/10.9734/AJEA/2013/2746.

[68] White, P. J., & Broadley, M. R. (2011). Physiological Limits to Zinc Biofortification of Edible Crops. *Frontiers in Plant Science*, *2*:80. https://doi.org/10.3389/fpls.2011.00080.

[69] Hunt, J.R. (2003). Bioavailability of Iron, Zink, and other trace minerals from vegetarian diets. *Am J Clin Nutr,* 78(3 Suppl):633S-639S.

[70] White, P. J., Broadley, M. R. (2009). Biofortification of crops with seven mineral elements often lacking in. *New Phytologist*, 182:49–84. https://doi.org/10.1111/j.1469-8137.2008.02738.x.

[71] Silva, P. S. C., Francisconi, L. S., Gonçalves, R. D. M. R. (2016). Evaluation of major and trace elements in medicinal plants. *Journal of the Brazilian Chemical Society*, 27(12): 2273–2289. https://doi.org/10.5935/0103-5053.20160123.

[72] Jabeen, S., Shah, M. T., Khan, S., Hayat, M. Q. (2010). Determination of major and trace elements in ten important folk therapeutic plants of Haripur basin, Pakistan. *Journal of Medicinal Plants Research*, 4(7):559–566.

[73] Weber, G., Konieczyński, P. (2003). Speciation of Mg, Mn and Zn in extracts of medicinal plants. *Analytical and Bioanalytical Chemistry*, 375(8):1067–1073. https://doi.org/10.1007/s00216-002-1706-z.

[74] Pizzorno, L. (2015). Nothing Boring About Boron. *Integrative Medicine (Encinitas, Calif.), 14*(4):35–48. https://doi.org/http://dx.doi.org/10.1016/j.blre.2012.12.003.

[75] Uluisik, I., Karakaya, H. C., Koc, A. (2018). The importance of boron in biological systems. *Journal of Trace Elements in Medicine and Biology*, 45:156–162. https://doi.org/10.1016/j.jtemb.2017.10.008.

[76] Welch, R. M., Shuman, L. (1995). Micronutrient Nutrition of Plants. *Critical Reviews in Plant Sciences,* 14(1):49-82. https://doi.org/10.1080/07352689509701922.

[77] Karak, T., Kutu, F. R., Nath, J. R., Sonar, I., Paul, R. K., Boruah, R. K., Sanyal, S., Sabhapondit, S., Dutta, A. K. (2017). Micronutrients (B, Co, Cu, Fe, Mn, Mo, and Zn) content in made tea (*Camellia sinensis* L.) and tea infusion with health prospect: A critical review. *Critical Reviews in Food Science and Nutrition*, 57(14):2996–3034. https://doi.org/10.1080/10408398.2015.1083534.

[78] Szentmihályi, K., Hajdú, M., Fodor, J., Kótai, L., Blázovics, A., Somogyi, A., Then, M. (2006). *In vitro* study of elements in herbal remedies. *Biological Trace Element Research*, 114(1–3):143–150. https://doi.org/10.1385/BTER:114:1:143.

[79] Yruela, I. (2005). Copper in plants. *Brazilian Journal of Plant*, 17(1):146–156. https://doi.org/10.1590/S1677-04202005000100012.

[80] Wrobel, J. K., Power, R., Toborek, M. (2016). Biological activity of selenium: Revisited. *IUBMB Life*, 68(2):97–105. https://doi.org/10.1002/iub.1466.

[81] White, P. J. (2015). Selenium accumulation by plants. *Annals of Botany*, 117(2):217-35. https://doi.org/10.1093/aob/mcv180.

[82] Kolbert, Z., Molnár, Á., Feigl, G., Van Hoewyk, D. (2019). Plant selenium toxicity: Proteome in the crosshairs. *Journal of Plant Physiology*, 232:291–300. https://doi.org/10.1016/j.jplph.2018.11.003.

[83] Mehdi, Y., Hornick, J. L., Istasse, L., Dufrasne, I. (2013). Selenium in the environment, metabolism and involvement in body functions. *Molecules*, 18(3):3292–3311. https://doi.org/10.3390/molecules18033292.

[84] Gupta, M., Gupta, S. (2017). An Overview of Selenium Uptake, Metabolism, and Toxicity in Plants. *Frontiers in Plant Science*, 7:2074. https://doi.org/10.3389/fpls.2016.02074.

[85] Sharma, S., Goyal, R., Sadana, U. S. (2014). Selenium Accumulation and Antioxidant Status of Rice Plants Grown on Seleniferous Soil from Northwestern India. *Rice Science*, 21(6): 327–334. https://doi.org/10.1016/S1672-6308(14)60270-5.

[86] Thosaikham, W., Jitmanee, K., Sittipout, R., Maneetong, S., Chantiratikul, A., Chantiratikul, P. (2014). Evaluation of selenium

species in selenium-enriched pakchoi (*Brassica chinensis* Jusl var parachinensis (Bailey) Tsen & Lee) using mixed ion-pair reversed phase HPLC–ICP-MS. *Food Chemistry*, *145*:736–742. https://doi.org/10.1016/j.foodchem.2013.08.116.

[87] Fraga, C. G. (2005). Relevance, essentiality and toxicity of trace elements in human health. *Molecular Aspects of Medicine*, 26(4–5):235–244. https://doi.org/10.1016/j.mam.2005.07.013.

[88] Weng, H.X., Liu, H.P., Li, D.W., Ye, M., Pan, L., Xia, T.H. (2014). An innovative approach for iodine supplementation using iodine-rich phytogenic food. *Environmental Geochemistry and Health*, 36(4):815–828. https://doi.org/10.1007/s10653-014-9597-4.

[89] World Health Organization and Food and Agriculture Organization of the United Nations. (2004). Vitamin and mineral requirements in human nutrition Second edition. *World Health Organization*, 1–20. https://doi.org/92 4 1546123.

[90] Madhusudan, C., Manoj, S., Rahul, K., Rishi, C. M. (2011). Seaweeds: A Diet with Nutritional, Medicinal and Industrial Value. *Research Journal of Medicinal Plant,* 5(2):153–157. https://doi.org/10.3923/rjmp.2011.153.157.

[91] Jabeen, S., Shah, M. T., Khan, S., Hayat, M. Q. (2010). Determination of major and trace elements in ten important folk therapeutic plants of Haripur basin, Pakistan. *Journal of Medicinal Plants Research*, 4(7):559–566.

[92] Demirezen, D., & Aksoy, A. (2006). Heavy metal levels in vegetables in Turkey are within safe limits for Cu, Zn, Ni and exceeded for Cd and Pb. *Journal of Food Quality*, 29(3):252–265. https://doi.org/10.1111/j.1745-4557.2006.00072.x.

[93] World Health Organization and Food and Agriculture Organization of the United Nations. (2004). Joint FAO/WHO Expert Consultation on Human Vitamin and Mineral Requirements. *Vitamin and mineral requirements in human nutrition.* 2nd edition. Bangkok, Thailand. http://apps.who.int/iris/bitstream/handle/10665/42716/9241546123.pdf;jsessionid=CD01E27C4F9AB7DE68920773B18B700E?sequence=1.

[94] Pu, F., Chen, N., Xue, S. (2016). Calcium intake, calcium homeostasis and health. *Food Science and Human Wellness*, 5:8-16.

[95] Beto, J.A. (2015). The role of calcium in human aging. *Clinical Nutrition Research*, 4:1-8.

[96] Paudel, S., Sindelar, R.,Saha, M. (2018). Calcium Signaling in Vertebrate Development and Its Role in Disease. *Int J MolSci*, 19(11). pii: E3390. doi: 10.3390/ijms19113390.

[97] European Food Safety Authority (EFSA). (2017). Dietary Reference Values for nutrients. *Summary report.* doi: 10.2903/sp.efsa.2017. e15121.

[98] Food and Nutrition Board, Institute of Medicine, National Academies. *Dietary Reference Intakes (DRIs): Recommended Dietary Allowances and Adequate Intakes, Elements.* https://ods. od.nih.gov/Health_Information/Dietary_Reference_Intakes.aspx.

[99] Schwalfenberg, G.K., Genuis, S.J. (2017). The Importance of Magnesium in Clinical Healthcare. *Scientifica* (Cairo), 2017: 4179326. doi: 10.1155/2017/4179326.

[100] de Baaij, J.H., Hoenderop, J.G.,Bindels, R.J. (2015). Magnesium in man: implications for health and disease. *Physiol Rev*,95(1):1-46. doi: 10.1152/physrev.00012.2014.

[101] Stone, M.S., Martyn, L., Weaver, C.M. (2016). Potassium intake, bioavailability, hypertension and glucose control. *Nutrients*, 8(7). pii: E444. doi: 10.3390/nu8070444.

[102] Sigel, A., Sigel, H., Siegel, R.K.O. (2013). Interrelations between essential metal ions and human diseases. Metal ions in life science. Chapter 2: Sodium and potassium in health and disease. *Springer Science and Business Media Dordrecht*, 13:49-79. doi: 10.1007/978-94-007-7500-8_3.

[103] Kovesdy, C.P., Appel, L.J., Grams, M.E., Gutekunst, L., McCullough, P.A., Palmer, B.F., Pitt, B., Sica, D.A., Townsend, R.R. (2017). Potassium homeostasis in health and disease: A scientific workshop cosponsored by the National Kidney Foundation and the American Society of Hypertension. *J Am Soc Hypertens*, 11(12):783-800. doi: 10.1016/j.jash.2017.09.011.

[104] United Nations Environment Programme, WHRC. (2007). *Reactive nitrogen in the environment. Too much or too little of a good thing.*

[105] Hord, N.G., Tang, Y., Bryan, N.S. (2009). Food sources of nitrates and nitrites: the physiologic context for potential health benefits. *Am J Clin Nutr*, 90(1):1-10.doi: 10.3945/ajcn.2008.27131.

[106] Raina, S., Garg, G., Sethi, S.K. et al. (2012). Phosphorus metabolism. *J Nephrol Therapeutic*, S_3:008. doi:10.4172/2161-0959.S3-008.

[107] Takeda, E., Taketani, Y., Sawada, N., Sato, T., Yamamoto, H. (2004). The regulation and function of phosphate in the human body. *Biofactors*, 21(1-4):345-55.

[108] Mukwevho, E., Ferreira, Z., Ayeleso, A. (2014). Potential role of sulfur-containing antioxidant systems in highly oxidative environments. *Molecules*, 19(12):19376-89. doi: 10.3390/molecules 191219376.

[109] Indian Agricultural Research Institute. (2014). Major sulphur compounds in plants and their role in human nutrition and health – an overview. *Proc Indian Natn Sci Acad,* 80(5):1045-1054.

[110] Wang, G., Nauseef, W.M. (2015). Salt, chloride, bleach and innate host defence. *J Leukoc Biol*, 98(2):163-172.

[111] Abbaspour, N., Hurell, R., Kelishadi, R. (2014). Review on iron and its importance for human health. *J Res Med Sci*, 19:164-74.

[112] Yokoi, K. (2018). Investigating the Essentiality and Requirements of Iron from the Ancient to the Present. *Biol Trace Elem Res*. doi: 10.1007/s12011-018-1584-7.

[113] Vidyavati, S.D., Sneha, A., Katti, S.M. (2016). Zinc: the importance in human life. *International J of Healthcare and Biomedical Research*, 4(4):18-20.

[114] Devi, C.B., Nadakishore, Th., Sangeeta, N. et al. (2014). Zinc in human health. *IOSR Journal of Dental and Medical Science*, 13(7):18-23.

[115] Vinceti, M., Filippini, T., Wise, L.A. (2018). Environmental Selenium and Human Health: an Update. *Curr Environ Health Rep*. doi: 10.1007/s40572-018-0213-0.

[116] Roman, M., Jitaru, P., Barbante, C. (2014). Selenium biochemistry and its role for human health. *Metallomics*, 6(1):25-54. doi: 10.1039/c3mt00185g.

[117] Velasco, I., Bath, S.C., Rayman, M.P. (2018) Iodineas Essential Nutrient during the First 1000 Days of Life. *Nutrients*, 10(3). pii: E290. doi: 10.3390/nu10030290.

[118] Khaliq, H., Juming, Z.,Ke-Mei, P. (2018). The Physiological Role of Borononon Health. *Biol Trace Elem Res,* 186(1):31-51. doi: 10.1007/s12011-018-1284-3.

[119] Nielsen, F.H. (2014). Update on human health effects of boron. *J Trace Elem Med Biol*, 28(4):383-7. doi: 10.1016/j.jtemb.2014.06.023.

[120] Bost, M., Houdart, S., Oberli, M., Kalonji, E., Huneau, J.F., Margaritis, I. (2016). Dietary copper and human health: Current evidence and unresolved issues. *J Trace Elem Med Biol*, 35:107-15. doi: 10.1016/j.jtemb.2016.02.006.

[121] Uriu-Adams, J.Y., Keen, C.L. (2005). Copper, oxidative stress and human health. *Molecular Aspects of Medicine*, 26:268-298.

[122] Osredkar, J., Sustar, N. (2011). Copper and zinc biological role and significance of copper/zinc imbalance. *J Clinic Toxicol*, S:3.

[123] Robinson, B.H. (1998). The role of manganese superoxide dismutase in health and disease. *J Inherit Metab Dis*, 21(5):598-603.

[124] Tuschl, K.,Mills, P.B.,Clayton, P.T. (2013). Manganese and the brain. *Int Rev Neurobiol*, 110:277-312. doi: 10.1016/B978-0-12-410502-7.00013-2.

[125] Lucchini, R., Placidi, D., Cagna, G., Fedrighi, C., Oppini, M., Peli, M., Zoni, S. (2017). Manganese and Developmental Neurotoxicity. *Adv Neurobiol*, 18:13-34. doi: 10.1007/978-3-319-60189-2_2.

[126] Zambelli, B., Ciurli, S. (2013). Nickel and human health. *Met Ions Life Sci*, 13:321-57. doi: 10.1007/978-94-007-7500-8_10.

[127] Zambelli, B., Uversky, V.N., Ciurli, S. (2016). Nickel impact on human health: An intrinsic disorder perspective. *Biochim Biophys Acta*, 1864(12):1714-1731. doi: 10.1016/j.bbapap.2016.09.008.

[128] Schwarz, G., Belaidi, A.A. (2013). Molybdenum in human health and disease. *Met Ions Life Sci*, 13:415-50. doi: 10.1007/978-94-007-7500-8_13.

[129] Ferreira, C.R., Gahl, W.A. (2017). Disorders of metal metabolism. *Transl Sci Rare Dis*, 18;2(3-4):101-139. doi: 10.3233/TRD-170015.

BIOGRAPHICAL SKETCH

Buda Valentina Oana

Affiliation: "Victor Babes" University of Medicine and Pharmacy, Department of Pharmacology and Clinical pharmacy

Education: "Victor Babes" University of Medicine and Pharmacy, Faculty of Pharmacy

Research and Professional Experience: University assistant (2010 – present). Clinical pharmacist (2015 – present). PhD thesis in Medicine – Title: The effects of Perindopril and Candesartan on the vascular endothelium in hipertensive patients, evaluated by the quantification of pentraxin-3, endoglin and thrombospondin-1

Publications from the Last 3 Years:

[1] Danciu C, Muntean D, Alexa E, Farcas C, Oprean C, Zupko I, Bor A, Minda D, Proks M, Buda V, Hancianu M, Cioanca O, Soica C, Popescu S, Dehelean CA. Phytochemical Characterization and Evaluation of the Antimicrobial, Antiproliferative and Pro-Apoptotic Potential of *Ephedra alata* Decne. Hydroalcoholic Extract against the MCF-7 Breast Cancer Cell Line. *Molecules.* 2018;24(1). pii: E13. doi: 10.3390/molecules24010013.ISI, impact factor 3.098/2017

[2] Buda V, Andor M, Ledeti A, Ledeti I, Vlase G, Vlase T, Cristescu C, Voicu M, Suciu L, Tomescu MC. Comparative Solid-State Stability of Perindopril Active Substance vs. Pharmaceutical Formulation. *Int J Mol Sci.* 2017; 15;18(1). doi: 10.3390/ijms 18010164. ISI, impact factor 3.687/2017.

[3] Buda V, Andor M, Petrescu L, Cristescu C, Baibata DE, Voicu M, Munteanu M, Citu I, Muntean C, Cretu O, Tomescu MC. Perindopril Induces TSP-1 Expression in Hypertensive Patients with Endothelial Dysfunction in Chronic Treatment. *Int J Mol Sci.* 2017; 18(2). doi: 10.3390/ijms18020348.ISI, impact factor 3.687/2017.

[4] Buda V, Andor M, Cristescu C, Voicu M, Cochera F, Tuduce P, Petrescu L, Tomescu MC. The effect of candesartan on pentraxin-3 plasma levels as marker of endothelial dysfunction in patients with essential arterial hypertension. *Ir J Med Sci.*2017; 186(3):621-629. doi: 10.1007/s11845-017-1580-5. ISI, impact factor 1.045/2017.

[5] Dehelean L, Andor M, Romosan AM, Manea MM, Romosan RS, Papava I, Bredicean AC, Buda VO, Tomescu MC. Pharmacological and disorder associated cardiovascular changes in patients with psychosis. A comparison between olazapine and risperidone. *Farmacia.* 2018; 66(1):129-134 ISI, impact factor 1.507/2017.

[6] Ledeti I, Pusztai AM, Muresan CM, Circioban D, Vlase G, Murariu MS, Suta LM, Vlase T, Ledeti A, Suciu O, Buda V, Matusz P. Study of solid-state degradation of prochlorperazine and promethazine. *Journal of Thermal Analysis and Calorimetry*. 2018. DOI: 10.1007/s10973-018-7566-3ISI, impact factor 2.209/2017.

[7] Ledeti I, Pusztai AM, Muresan CM, Circioban D, Vlase G, Axente C, Suta LM, Vlase T, Matusz P, Ledeti A, Suciu O, Buda V, Murariu M. Study of kinetic characteristics of 6-ethyl chenodeoxycholic acid and deoxycholic acid decomposition in non-isothermal conditions. *Journal of Thermal Analysis and*

Calorimetry. 2018. DOI:10.1007/s10973-018-7660-6 ISI, impact factor 2.209/2017.

[8] Cristescu C, Negres S, Suciu M, Voicu A, Buda V, Suciu L, Proks M, Voicu M. Study regarding the parents' use of self-medication among children under 12 years old. *Farmacia.* 2018; 66(5):811-819. DOI:10.31925/farmacia.2018.5.10 ISI, impact factor 1.507/2017.

[9] Buda V, Andor M, Cristescu C, Voicu M, Suciu L, Muntean C, Cretu O, Baibata DE, Gheorghiu CM, Tomescu MC. The influence of perindopril on PTX3 plasma levels in hypertensive patients with endothelial dysfunction. *Farmacia.* 2016; 64(3):382-389ISI, impact factor 1.348/2016.

[10] Suciu L, Cristescu C, Topirceanu A, Udrescu L, Udrescu M, Buda V, Tomescu MC. Evaluation of patients diagnosed with essential arterial hypertension through network analysis. *Ir J Med Sci.*2016; 185(2):443-51. doi: 10.1007/s11845-015-1342-1ISI, impact factor 1.224/2016.

In: Phytochemicals
Editor: Iman Ryan
ISBN: 978-1-53615-478-8

Chapter 2

DIETARY SOYBEAN ISOFLAVONES IN LIFESTYLE-RELATED DISEASE PREVENTION

Kazuo Yamagata**[*], **PhD
Department of Food Bioscience & Biotechnology,
College of Bioresource Science, Nihon University (NUBS),
Fujisawa, Kanagawa, Japan

ABSTRACT

Soybeans are among the most popular foods worldwide and consumption of soy-containing foods has been associated with health benefits. Soybeans contain saponins, soy proteins and peptides, and flavonoids, all of which exert multiple biological activities. In particular, numerous epidemiologic studies have indicated an inverse relationship between the consumption of soy isoflavones and the incidence of lifestyle-related diseases. For example, soy isoflavones reduce plasma total cholesterol and triglycerides and increase plasma leptin levels. In addition, consumption of isoflavones is associated with a lower risk of type 2 diabetes in post-menopausal women. Epidemiological studies also indicate that high consumption of soy isoflavones is associated with a lower incidence of hormone-dependent cancers, such as breast and

[*] Corresponding Author's E-mail: kyamagat@brs.nihon-u.ac.jp.

prostate cancer. In ovariectomized mice, isoflavone-enriched dietary supplementation prevents bone mineral density reduction and enhances bone formation. However, although the evidence supporting the benefits of soy phytochemicals in disease prevention continues to increase, the association between soy isoflavones and disease is not fully understood. Typical soy isoflavones include daidzein and genistein. Furthermore, equol, a major metabolite of daidzein, is generated by enterobacterial effects. This review provides an overview of the recent progress in identifying the mechanisms of action underlying the potential roles of dietary soybean isoflavones including daidzein, genistein, and equol that may have benefits in dyslipidemia, ischemic heart disease, blood pressure, diabetes, cancer, and osteoporosis.

Keywords: soy isoflavone, disease prevention, cholesterol reduction, dietary supplementation, diadzein

ABBREVIATIONS

ACE	angiotensin-converting enzyme
AVP	arginine vasopressin
CRP	C-reactive protein
ER	estrogen receptor
GPx	glutathione peroxidase
I/R	ischemia/reperfusion
IL-6	interleukin 6
LDL	low-density lipoprotein
MCAO	middle cerebral artery occlusion
miRNA	(miR), microRNA
ROS	reactive oxygen species
SHRSP	stroke-prone spontaneously hypertensive rats
SOD	superoxide dismutase
TNFα	tumor necrosis factor-alpha
TGF-β1	transforming growth factor beta 1

1. INTRODUCTION

Soybeans constitute one of the world's most popular foods, and numerous studies have indicated an inverse relationship between the consumption of soy isoflavones and the incidence of chronic diseases such as life-style-related diseases (Omoni and Aluko, 2005). Soybeans contain saponins, soy proteins and peptides, and flavonoids, all of which exert multiple biological activities. For example, the preventative effects of soy intake on atherosclerosis are considered to be mediated primarily by the phytoestrogen component (Anthony et al., 1997). Furthermore, studies have indicated the beneficial effects of several soy isoflavones against hypercholesterol (Hall et al., 2006; Zung et al., 2010) as well as the ability of isoflavonoids (Schabath et al., 2005; Andres et al., 2011) and saponins (Tsai et al., 2010) to decrease cancer risks.

Soy isoflavones comprise polyphenolic compounds that exert weak estrogenic or hormone-like activity via structural similarity with 17-β-estradiol. In particular, isoflavones induce estrogen-agonist and/or estrogen-antagonist activities (Lamp, 2003). Estrogens bind to estrogen receptors (ERs) within cells and subsequently enhance signaling molecules to exert their effects. The estrogen-mimicking effects of isoflavones regulate several pathological conditions, such as cancer, metabolic and cardiovascular diseases, inflammation, and osteoporosis (Paterni et al., 2014). Soy isoflavones have also been shown to inhibit adipogenesis and lipogenesis by interacting with transcription factors and upstream signaling molecules (Zhang et al., 2009). Furthermore, soy isoflavones inhibit myocardial ischemia/reperfusion (I/R) injury by enhancing endothelial nitric oxide synthase (eNOS) (Tang et al., 2016), and also reduce myocardial damage (Kimet al., 2009). The anti-ischemic heart disease effects of soy isoflavones include a marked induction of superoxide dismutase and glutathione peroxidase (GPx) activities and a reduction in the level of malondialdehyde in endothelial cells (Jin et al., 2015). These effects contribute to the mechanisms underlying the health benefits of soy isoflavones. Moreover, enhanced isoflavone intake is associated with a lower risk of type 2 diabetes in post-menopausal women (Ding et al., 2015) along with a lower incidence of hormone-dependent cancers including breast and prostate cancer (Wada et al., 2013; Applegate et al., 2018).

Recognition of the benefits of soy in disease prevention continues to increase; however, the association between soy isoflavones and disease is not fully understood. The typical isoflavones in soybean comprise genistein and daidzein, with equol, a major metabolite of daidzein, being generated by enterobacterial effects. This review provides an overview of the recent progress in identifying the mechanisms of action underlying the potential roles of dietary soy isoflavones that may have benefits in dyslipidemia, ischemic heart disease, blood pressure, diabetes, cancer, and osteoporosis.

2. Nutritional Components of Soybean

Soybeans contain considerable amounts of protein, carbohydrate, and lipids (Mateos-Aparicio et al., 2008). In particular, whereas the protein content of most beans averages 20–25%, soybeans contain 30–40% protein (Table 1) (Mateos-Aparicio et al., 2008).

Moreover, soybean protein only contains a small percentage of sulfur-containing amino acids, particularly methionine, with limited cystine and threonine. As a whole, the amino acid content of soybean protein is equivalent to that of most animal protein (Mateos-Aparicio et al., 2008).

As shown in Table 1, soybeans also contain approximately 30 g carbohydrates per 100 g dry matter. Carbohydrate content thus represents the second largest component in soybeans, with the main monomers of polysaccharides in soybean being glucose, galactose, uronic acids, arabinose, and xylose (Redondo-Cuenca et al., 2008).

In addition, soybeans contain 18–22% lipids, of which 99% are triglycerides. Other lipid components include unsaponifiable matter such as tocopherols, phytosterols, and carbohydrates, along with free fatty acids, the majority of which comprise polyunsaturated fatty acids (e.g., linoleic and linolenic acids). Unsaturated fatty acids such as oleic acids and saturated fatty acids including palmitic and stearic acids are also present. Among these, the content of linoleate is the highest, representing approximately 7–8% of the total nutritional content of the soybean (Messina, 1999).

3. ISOFLAVONES

The activity of soybean to inhibit chronic disease has been previously indicated to be present in trace constituents including polyphenol, isoflavones, and dietary fiber. In particular, isoflavones, which comprise non-nutritive substances that are classified as a subgroup of polyphenol flavonoids, are considered to strongly contribute to the disease prevention capabilities of soybeans as they possess health protective effects against several diseases (Gil-Izquierdo et al., 2012). Specifically, isoflavones contribute to the prevention and treatment of chronic diseases including heart disease, cancer, diabetes, and hypertension (Messina, 1999).

Table 1. Nutritional content of soybeans (100 g dry matter)

	Composition	Content (g)	Total content (g)
Protein		36	36
Carbohydrates	Complex carbohydrates Simple carbohydrates	21 9	30
Fat	Saturated fat Monounsaturated fat Polyunsaturated fat	2.8 4.4 11.2	19
Fiber	Insolube 10 Solube 7	10 7	17
Minerals	Calcium Magnesium Potassium Iron Zinc	0.276 0.280 1.797 0.016 0.0048	2.4

Adapted from Mateos-Aparicio, et al., (2008).

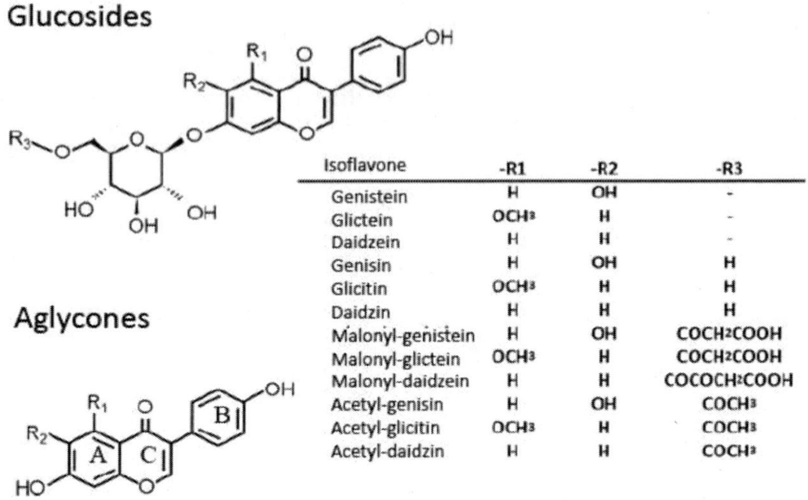

Isoflavone	-R1	-R2	-R3
Genistein	H	OH	-
Glictein	OCH_3	H	-
Daidzein	H	H	-
Genisin	H	OH	H
Glicitin	OCH_3	H	H
Daidzin	H	H	H
Malonyl-genistein	H	OH	$COCH_2COOH$
Malonyl-glictein	OCH_3	H	$COCH_2COOH$
Malonyl-daidzein	H	H	$COCOCH_2COOH$
Acetyl-genisin	H	OH	$COCH_3$
Acetyl-glicitin	OCH_3	H	$COCH_3$
Acetyl-daidzin	H	H	$COCH_3$

Adapted from Vitale, D.C., et al., (2013).

Figure 1. Chemical structure of glycoside and aglycone isoflavones.

Table 2. Estimated isoflavone contents of selected soy protein products

Soy protein product	Daidzein	Genistein	Glycitein	Total isoflavones
Soy flour, full-fat, roasted	99.27	98.75	16.40	198.95
Soybean flakes, defatted	36.97	85.69	14.23	125.82
Soy protein concentrate, aqueous washed	43.04	55.59	5.16	102.07
Soy protein concentrate, produced by alcohol extraction	6.83	5.33	1.57	12.47
Soy protein isolate	33.59	59.62	9.47	97.43

Values based on mg/100 g edible portion.

Adapted from Erdman, et al., (2004).

3.1. Chemical Structure of Isoflavones

The basic chemical structure of flavonoid compounds is the flavone nucleus, which has two benzene rings (A and B) linked with heterocyclic pyrane (Figure 1) (Vitale et al., 2013). As shown in Figure 1, several isoflavones including aglycones and glycosides can be identified based on this structure, with typical isoflavones in soybean being genistein and daidzein. Furthermore, the intestinal bacteria can convert daidzein into several metabolites including the soflavonoids equol, dihydrodaidzein, and O-desmethylangolensin (Joannou et al., 1995). Notably, the isoflavones in nonfermented soyfoods are mostly the conjugated type, whereas in fermented soy food products such as miso, aglycones are more common (Wang et al., 1994).

3.2. Contents of Isoflavones in Soy and Soy Foods

Soy isoflavones are included in soy, soy foods, and legumes. In soybean foods, isoflavone content is reported as 0.1–3.0 mg/g (Mateos-Aparicio et al., 2008). Table 2 indicates the total and individual isoflavone contents of several soy foods (Erdman et al., 2004). In particular, roasted soy flour has a large isoflavone content, with a high content of daidzein and genistein. Notably, however, the isoflavone content and composition is markedly changed by processing (Ryu et al., 2010). For example, the glycoside of miso and soy sauce is converted into aglucon by the effects of microbial α-glucosidase in the fermentation process (Toda et al., 1999). In contrast, in the natto produced with these fermented foods, the aglucon content does not increase.

4. Protective Effects of Isoflavones against Lifestyle-Related Disease

Lifestyle, including diet, is associated with many diseases including diabetes, hyperlipidemia, hypertension, and hyperuricemia. Among these, heart disease, stroke, obesity, and type 2 diabetes comprise are particularly important (Haskell, 2003). In this section, the effects of isoflavones on various lifestyle diseases are described.

4.1. Dyslipidemia

Epidemiologic study has indicated that the dietary consumption of phytoestrogens was associated with a favorable metabolic cardiovascular risk profile in 939 postmenopausal women in the United States (de Kleijn et al., 2002). Furthermore, in rats fed a high cholesterol diet, the provision of isoflavone aglycones in lactic acid-fermented soy milk improved lipid metabolism (Kobayashi et al., 2014). These results suggest that the amelioration of lipid-associated disease risk via enhanced lipid metabolism is synergistic with soybean protein and isoflavone ingestion.

4.1.1. Genistein

In comparison, a study examining the effect of genistein on high fat diet in male hamsters found that genistein intake reduced high fat diet-induced plasma cholesterol, triglycerides, low-density lipoprotein (LDL) cholesterol, and malondialdehyde (Tang et al., 2015). In addition, genistein enhanced hepatic LDL receptor and ERα expression. These results suggest the beneficial effects of genistein on hyperlipidemia.

4.1.2. Daidzein

Equol constitutes a major isoflavone metabolite from daidzein and exhibits estrogen-like biological activity (Birru et al., 2016). Equol has been shown to effect the plasma lipid profile. For example, a human study including men with hypercholesterolemia (85 men) and postmenopausal women (among 42 men and 43 women) studied the effects of soy food ingestion over 1 month at doses of 30–52 g/day for three conditions: (1) without isoflavones, (2) 10 mg/day, and (3) 73 mg/day (Wong et al., 2012). The percentage changes from baseline in equol producers are shown in Figure 2 (Wong et al., 2012). Soy foods decreased serum LDL cholesterol equally in equol producers and nonproducers. In contrast, in equol producers, soy consumption supported the maintenance of higher high-density lipoprotein (HDL)-cholesterol concentrations than those seen in equol nonproducers. These data indicate that soy foods led to significant reductions in LDL cholesterol and apo B in equol producers.

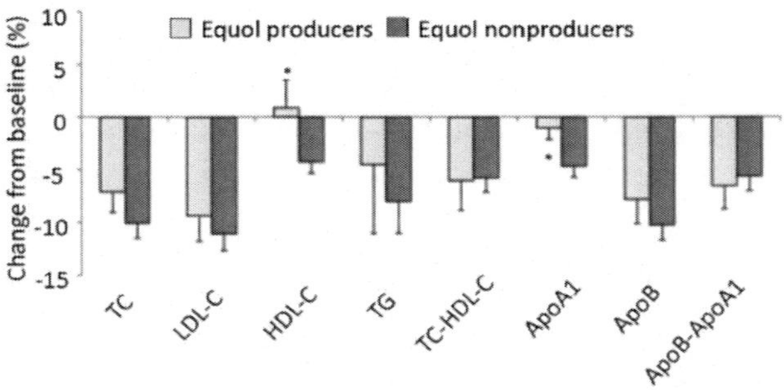

n = 85; 30 equol producers, 55 nonproducers. *Significantly different from equol nonproducers assessed by using ANCOVA, with mean change from baseline as the response variable and study, equol status, and study * equol status interaction as the main effects (P <, 0.05). ApoA1, apolipoprotein A-I; ApoB, apolipoprotein B; HDL-C, HDL cholesterol; LDL-C, LDL cholesterol; TC, total cholesterol; TG, triglycerides.

Adapted from Wong, J.M., et al. (2012).

Figure 2. Changes in serum lipids and apolipoproteins in equol producers and nonproducers upon soy ingestion.

4.2. Heart Disease (Particularly Ischemic Heart Disease)

Soybean phytoestrogens affect coronary arteries in patients with ischemic heart disease and prevent cardiac disease. Specifically, higher serum genistein levels are associated with endothelial-dependent coronary microvascular function in women with suspected myocardial ischemia (Pepine et al., 2007). In addition, isoflavone intake decreases the levels of triglyceride and LDL, and enhances brachial flow-mediated dilatation in subjects (Li and Zhang, 2017). Simultaneously, isoflavone increases Nrf2 and superoxide dismutase (SOD), and reduces C-reactive protein (CRP), malondialdehyde, 8-isoprostane, tumor necrosis factor-alpha (TNFα), and interleukin-6 (IL-6) in serum. Moreover, the effects of isoflavone are closely related to a decrease of flow-mediated dilatation impairment. Together, these results indicate that the effect of soybean isoflavones on the amelioration of ischemic cardiomyopathy is caused in part by Nrf2-mediated activation of the antioxidant response.

4.2.1. Genistein

Similarly, genistein prevents diabetic myocardium impairment via anti-inflammatory and antioxidant effects in streptozotocin-induced diabetic cardiomyopathy in rats (Gupta et al., 2015). Specifically, genistein reduced streptozotocin-induced increases in blood glucose, HbA1c, CRP, transforming growth factor beta 1 (TGF-β1), and TNF-α, and enhanced the antioxidant activity of the myocardium.

The mechanisms underlying effects of genistein on myocardial fibrosis have been also examined in type 1 diabetic rats (Yang et al., 2018). Genistein reduced myocardial fibrosis in type 1 diabetic rats concomitant with inhibition of the inflammatory reaction along with the TGF-β1/Smad3 signaling pathway to regulate collagen production. These results suggest that the preventive mechanisms of genistein may be associated with the decrease of creatine kinase MB isozyme (CK-MB) and lactate dehydrogenase (LDH) release in diabetic rats. Furthermore, genistein inhibited isoproterenol-enhanced cardiac hypertrophy in male Wistar rats (Maulik et al., 2012). Genistein prevented isoproterenol-enhanced heart weight, left ventricular mass, myocardial 1-OH proline, fibrosis, myocyte size, and myocardial oxidative stress. Conversely, these varied effects of genistein were inhibited by the NOS inhibitor L-NAME. The effects of genistein on isoproterenol-enhanced cardiac hypertrophy suggest that they may be mediated through the enhancement of eNOS activities.

In addition, genistein has also been shown to have an effect on the heart damage induced by (I/R in rabbits (Ji et al., 2004). In rabbit coronary arteries, 45 min heart ischemia and 180 min reperfusion induced increased infarction size and apoptosis. However, genistein decreased the infarct size and apoptosis of myocytes upon I/R in rabbit heart. In cardiomyoblast cells, isoproterenol induced apoptosis and enhanced Bad, caspase-8, -9, and -3 proteins (Hu et al., 2013). In contrast, genistein inhibited the isoproterenol-induced apoptosis-related protein expression. Further analysis demonstrated that the genistein-mediated blockage of apoptosis involved the p38, Erk1/2, JNK, and NFκB signaling pathways (Hu et al., 2013).

4.2.2. Daidzein

In a rat I/R model, daidzein reduced myocardial damage and infarction size via the decrease of cytokine release and nuclear translocation of NF-κB (Kim et al., 2008). Furthermore, daidzein decreased plasma TNF-α and IL-6 levels and reduced malondialdehyde levels, myeloperoxidase activity, catalase activity, and neutrophil infiltration of rat myocardium. Thus, daidzein appeared to prevent I/R-induced myocardial damage through inhibition of NF-κB activation, and potentially blocked inflammatory cytokine expression. In addition, in cultured cardiomyoblast cells, daidzein also inhibited isoproterenol-induced apoptosis, similar to the effects of genistein (Hu et al., 2016). However, the effect of daidzein was caused by Akt pathway signaling. Furthermore, in diabetic cardiomyopathy, a glucose toxicity-enhanced cardiac mechanical malfunction (Hintz and Ren, 2004), daidzein prevented glucose-mediated myocyte dysfunction, suggesting a preventive activity against diabetes-associated cardiac defects.

4.3. Blood Pressure

In 40 men and women with mild-to-moderate hypertension. soy milk intake induced reduced systolic blood pressure (Rivas et al., 2002). Moreover, the hypotensive action of soy milk was correlated with the urinary excretion of genistein. These results indicated that chronic soy milk intake exerted hypotensive action in subjects with essential hypertension. The effects of genistein on blood pressure and ultrastructural changes in the kidney have also been investigated in fructose-fed hypertensive rats (Palanisamy and Venkataraman, 2013). Fructose-feeding enhanced blood pressure and heart rate in hypertensive rats. In contrast, genistein (1 mg/kg per day) intake induced reduced blood pressure and restored plasma angiotensin-converting enzyme (ACE) activity, along with protein kinase C-βII (PKC-βII) and eNOS expression. Genistein intake also preserved renal ultrastructural integrity. In particular, genistein influences eNOS activation and NO in renal cells. In comparison, in overweight men and postmenopausal women (25 subjects), a metabolite of daidzein (transtetrahydrodaidzein) reduced systolic blood pressure and cardiovascular risk (Nestel et al., 2007). Such metabolites normally form after consumption of isoflavones. Daidzein metabolites, therefore, may also have the ability to reduce blood pressure.

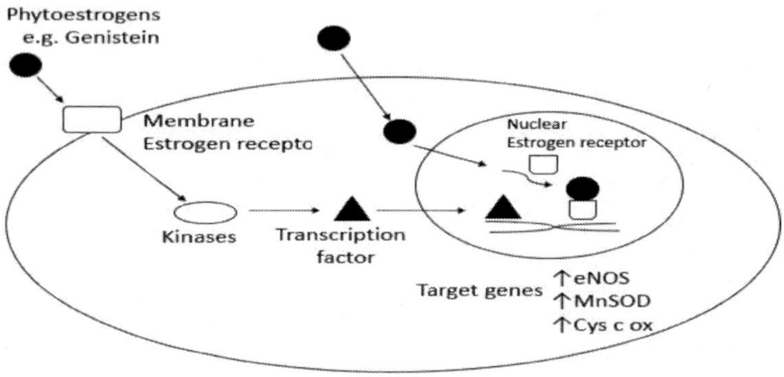

Activation of membrane estrogen receptors (ER) can trigger signaling cascades leading to increased eNOS activity and the induction of antioxidant defense genes via kinase mediated modulation of transcription factors. Classical nuclear targets for estrogen and phytoestrogens (e.g., genistein) can modulate eNOS and antioxidant gene expression through the interaction of receptor complexes with the *ERE* and/or *antioxidant response element* (*ARE*) in the promoter region of target genes. cytochrome *c*oxidase (Cys c ox).

Adapted from Mahn, K., et al., (2005).

Figure 3. Dietary soy isoflavones increase eNOS and antioxidant gene expression.

Furthermore, a study indicated that rats fed with a soy protein-rich diet during gestation and adult life resulted in the reduction of oxidative stress (Mahn et al., 2005). These animals also demonstrated improved endothelial function and decreased blood pressure in aged male offspring. The soy protein-rich diet appeared to reverse the aging-association alteration in mitochondrial glutathione, NOS expression, and manganese superoxide dismutase (MnSOD). In contrast, the reduced eNOS expression and antioxidant expression in aging animals induced impaired endothelial function and elevated blood pressure. These findings suggest that the ingestion of soy isoflavones would lead to an improvement in endothelial function, increased gene expression, and reduced blood pressure in postmenopausal women and patients at risk of coronary heart disease. As shown in Figure 3, estrogens and phytoestrogens enhance transcriptional activation of intracellular ERs (Mahn et al., 2005), which subsequently increase the expression of genes including those coding for eNOS, and MnSOD. In a similar manner, dietary soy isoflavones are also able to increase eNOS and antioxidant gene expression (Figure 3).

4.3.1. Genistein

Genistein has been shown to prevent the hypertensive effects of NaCl intake in male stroke-prone spontaneously hypertensive rats (SHRSP) (Cho et al., 2007). Furthermore, in SHRSPs fed a high NaCl diet, genistein decreased plasma insulin and insulin resistance and reduced plasma cholesterol and triglycerides. Genistein (1–10 μM) also enhanced the expression of human eNOS and reduced blood pressure in spontaneously hypertensive rats (Si and Liu, 2008). Effects of genistein on eNOS and NO were not with inhibiting of tyrosine kinases or activation of estrogen signaling. Furthermore, genistein reduced ACE and improved the vascular responses to angiotensin I and bradykinin (Montenegro et al., 2009). These results indicate that the inhibition of ACE along with antihypertensive effects enhance the beneficial vascular effects produced by genistein. Moreover, genistein affected the expression of genes such as those encoding endothelin-converting enzyme-1, endothelin-2, estrogen-related receptor alpha, and atrial natriuretic peptide receptor A precursor associated with vascular tone (Ambra et al., 2005). Similarly, genistein related the effect of oxidized (ox) LDL on the levels of mRNA encoding vascular endothelial growth factor receptor 165, types 1 and 2. These findings indicate that modulation of gene expression by genistein underlies its association with blood pressure regulation in addition to the reduction of angiogenic response to oxLDL in endothelial cells.

In contrast, in rabbit aortic smooth muscle, genistein induced the dose-dependent relaxation of 40 mM KCl-precontracted arterial strips, and reduced the contractile responses of noradrenaline, KCl, and $CaCl_2$ (Li et al., 2004). However, the vasodilator effect of genistein was endothelium independent and not associated with NO. This report suggested that genistein inhibits Ca^{2+} influx by calcium channels in a noncompetitive manner and prevents the noradrenaline-enhanced Ca^{2+} release from intracellular stores. Conversely, the detailed mechanism underlying the effects of genistein with regard to NO production remain unknown, although the lack of association with vasodilator effect may be related to a difference in experimental conditions.

4.3.2. Daidzein

Daidzein is reported to reduce blood pressure and artery smooth muscle cell contraction by inhibiting the receptor-mediated Ca^{2+}-influx in SHRs (Cao et al., 2006). Furthermore, a metabolite of the isoflavone daidzein, equol, has been associated with beneficial health effects. In postmenopausal women taking tibolone (synthetic estrogen), equol consumption reduced blood pressure compared to that in non-equol producers (Tormala et al., 2007).

4.4. Diabetes Mellitus

Several studies have demonstrated the relationship between soy consumption and type 2 diabetes mellitus risk. Recently, in a meta-analysis including 8 studies with 19 reports, inverse association was indicated between soy consumption and type 2 diabetes mellitus risk (Li et al., 2018). The association between soy food and isoflavone intake and risk of type 2 diabetes was also investigated in men and women in the United States (Ding et al., 2016). From 1966 including over 321 person-years of follow-up, 9,185 incidences of type 2 diabetes were identified. The results indicated that the consumption of isoflavones such as daidzein and genistein had inverse association with the risk of type 2 diabetes. Motrobrt, an *in vitro* study indicated that the isoflavones in chickpeas induced glucose uptake and glucose transporter 4 (Glut4) expression in 3T3-L1 cells (Gao et al., 2015). Together, these results indicate that soy protein and soy isoflavones may be associated with a decreased risk of type 2 diabetes mellitus.

4.4.1. Genistein

Genistein and soy protein supplementation enhanced the plasma insulin level and reduced the HbAIC level of streptozotocin-induced male diabetic rats (Lee, 2006). Furthermore, genistein and soy protein decreased SOD, catalase, and GPx activities in hepatocytes. Hyperglycemia-induced oxidative stress has been related in the development of diabetic nephropathy.

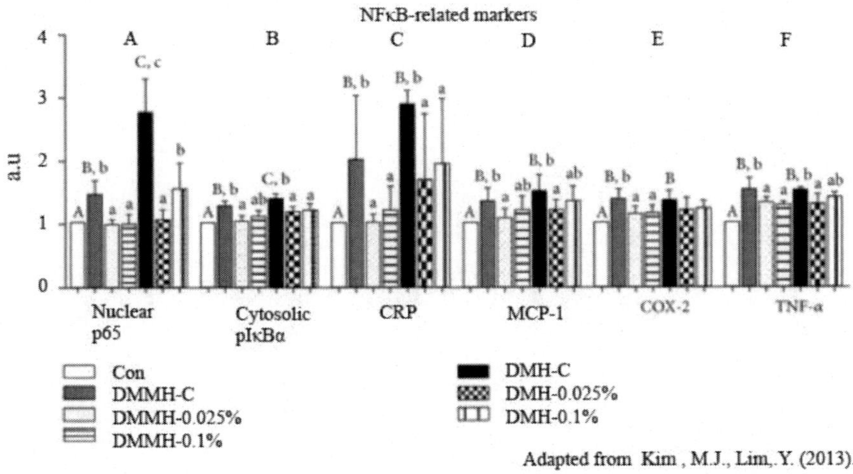

Effect of genistein supplementation on the kidney protein levels of Mcytosolic Nrf2 (A), HO-1 (B), GPx (C), CuZnSOD (D), COX-2 (E), and MnZnSOD (F) in experimental mice. All results were conducted at least three times. Data are presented as means SD (-10/group). Mean values with different letters were significantly different. Statistical differences of variables among CON, DMMH-C, and DMH-C analyzed by one-way ANOVA were shown in capital letters and effects of the genistein supplemented diet and/or DM severity using two-way ANOVA were represented by small letters.

Adapted from Kim, M.J., Lim., Y. (2013).

Figure 4. Effect of genistein supplementation on the kidney protein levels of Nrf2-related markers.

In human, genistein exhibited protective effects against high glucose-enhanced cell apoptosis in pancreas cells (Zhong et al., 2011). In particular, these effects were mediated via ER β and Bcl-2 signaling pathways. In addition, in β-cells of streptozotocin-induced diabetic rats, genistein and high-dose genistein improved the morphology of islets with enhanced insulin immunoreaction (El-Kordy and Alshahrani, 2015). Genistein also inhibited β-cell loss and improved glucose and insulin levels. These results suggest that genistein has a protective effect against pancreatic β-cell lesions. In particular, genistein possesses the ability to regenerate β-cells and improve serum levels of insulin in streptozotocin-induced diabetic rats.

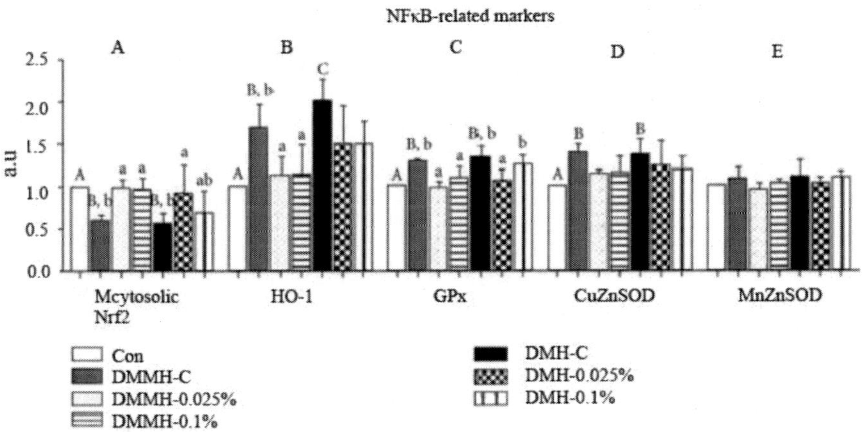

Effect of genistein supplementation on the kidney protein levels of p65 (NFκB) (A), pIκBα (B), CRP (C), MCP-1 (D), COX-2 (E), and TNF-α (F) in experimental mice. All results were conducted at least three times. Data are presented as means ± SD (n = 9-10/group). Mean values with different letters were significantly different, $P < 0.05$. Statistical differences of variables among CON, DMMH-C, and DMH-C analyzed by one-way ANOVA were shown in capital letters and effects of the genistein supplemented diet and/or DM severity using two-way ANOVA were represented by small letters.

Adapted from Kim, M.J., Lim., Y. (2013).

Figure 5. Effect of genistein supplementation on the kidney protein levels of p65 (NFκB), pIκBα, CRP, MCP-1, COX-2, and TNF-α in experimental mice.

Genistein supplementation also prevented levels of fasting blood glucose in the diabetic groups in mouse (Kim and Lim, 2013). As shown in Figure 4 and 5, genistein intake enhanced the expression of anti-oxidant enzymes such as heme oxygenase-1 (HO-1), GPx, and SOD, and reduced inflammatory markers such as nuclear factor kappa B (p65), phosphorylated inhibitory kappa B alpha (IkB), CRP, monocyte chemotactic protein-1 (MCP-1), cyclooxygenase-2 (COX-2), and TNF-α.

Soybean isoflavones induced the expression of the SHARP-2 component of the insulin-dependent signaling pathway associated with the regulation of blood glucose in H4IIE cells (Haneishi et al., 2011).

In addition, it was demonstrated that genistein inhibited features of diabetes-enhanced vascular inflammation, such as monocyte adhesion to human aortic endothelial cells, which may be mediated by promoting the cAMP/PKA pathway (Babu et al., 2012). Furthermore, genistein reduced the high glucose-induced gene expression of TGF- β2, alpha B-crystallin, fibronectin, and reduced glutathione levels in HLE-B3 cells. In turn, these effects of genistein may result in the inhibition of aldose reductase and cellular oxidation. As aldose reductase is involved in the development of diabetic complications including cataracts (Kim et al., 2008), these findings suggest that genistein may be effective for the prevention of diabetic complications such as diabetic cataracts.

4.4.2. Daidzein

Daidzein prevents the altered glucose and abnormal lipid metabolism and vascular inflammation associated with type 2 diabetes (Das et al., 2018). Daidzein inhibited the rise in fasting blood glucose and induced glucose uptake and GLUT4 translocation by AMPK phosphorylation (Cheong et al., 2014). In addition, the equol produced from daidzein via gut microbiota has been implicated in preventing type 2 diabetes mellitus (Horiuchi et al., 2017). Specifically, equol induced glucose-enhanced insulin secretion in the pancreatic islets of mouse, and may increase murine β-cell function and inhibit hypoglycemia. Thus, daidzein may be useful for the prevention of diabetes.

4.5. Breast Cancer

Breast cancer is one of the most common cancers in women. For example, in the United States, breast cancer constitutes approximately 15% of all female cancer-related deaths (Ferlay et al., 2015). In contrast, an epidemiologic study showed that soy food intake prevents cancers such as breast and prostate cancer (Tse and Eslick et al., 2016). In particular, the effects of the isoflavone on cancer cells contribute to the prevention of cancer (Ko, 2014). The anti-proliferative effects of soy isoflavones on cancer cells has been associated with their interactions with the ER (Uifalean et al. 2015). These effects augment apoptoses, inhibit angiogenesis and metastasis, and regulate cell proliferation and survival. This section describes the protective effects against breast and prostate cancer of the isoflavonoids genistein and daidzein.

4.5.1. Genistein

Genistein was originally demonstrated as an inhibitor of tyrosine kinases (Akiyama et al., 1987) and topoisomerase (Corbett et al., 1993), and was also shown to bind to the ER (Wang et al., 1996). Furthermore, genistein regulates a number of important intracellular molecules, such as NF-κB and MAPK, to enhanced growth blockage and apoptosis inhibition. Such genistein functions may have anti-cancer effects.

Genistein induces the proapoptotic BAX/Bcl-2 ratio and enhances the reduction of antiapoptotic survivin in MCF-7 cells (Prietsch et al., 2014). Moreover, genistein increases the gene expression of antioxidant enzymes including MnSOD, CuZnSOD, thioredoxin reductase (TrxR), and GPx. Furthermore, genistein induces phosphatidylserine externalization and LC3A/B immunopositivity of autophagy-related proteins. In addition, genistein inhibits the proliferation of MCF-7 cells and enhances cell apoptosis by blocking IGF-1R-PI3 K/Akt signaling and reducing Bcl-2/Bax expression (Chen et al., 2015). In breast cancer cells, genistein induces apoptosis and growth inhibition (Zhao et al., 2016). These effects of genistein correlate with the presence of protein phosphatase 2A (CIP2A) (Tseng et al., 2012), a novel oncogene overexpressed in breast cancer.

Moreover, genistein inhibits oxidative stress and prevents mitochondrial function through the regulation of antioxidant enzymes in breast cancer MCF-7 cells via the ERα/ERβ ratio (Nadal-Serrano et al., 2013). These results indicate that the anti-cancer effect of genistein involves antioxidant enzymes and apoptotic signaling induction, and results in apoptosis and the induction of autophagy.

Furthermore, it has also been shown that genistein induces p53 protein, phosphorylation of p53, and phosphorylation of the hCds1/Chk2 protein kinase (Ye et al., 2000). These data indicate that genistein may serve as a key activating factor of a DNA damage-induced pathway that activates p53 and Chk2. In addition, genistein exerts preventative effects against 2,3,7,8-tetrachlorodibenzo-p-dioxin (TCDD)-induced reduction in BRCA-1 expression in MCF-7 cells (Romagnolo et al., 2017). In MCF-7 cells, daidzein and genistein exhibit biphasic effects (stimulatory or inhibitory) on cell growth and ERα expression (Choi and Kim, 2013). At 1 μM, daidzein induces cell growth but does not affect ER expression. In contrast, exposure to <10 μM genistein enhances cell growth and ERα expression whereas at 200 μM, genistein-mediated inhibition of cell growth markedly increases as compared that of daidzein. Genistein also reduced DNA methyltransferase (DNMT), and cyclin D1 in MCF-7 cells, thereby contributing to BRCA1 CpG methylation and further downregulation in ERα-positive breast cancer cells.

Notably, genistein also reduced multidrug resistance in MCF-7 cells through the translational regulation of ABCG2 transporters with reduced microRNA (miR)-181 (Rigalli et al., 2016). The loss of multidrug resistance consequent to reduced expression of the ABC drug transporter leads to the success of chemotherapy. Therefore, genistein treatment may be beneficial for patients with tumors expressing ABCG2.

Moreover, gene microarray data demonstrated that genistein stimulated expression changes of differentially expressed genes (DEGs) in breast cancer cell lines (Zhang et al., 2014). These genes included several cell cycle-related genes such as *GLIPR1*, *CDC20*, *BUB1*, *MCM2*, and *CCNB1*. Consistent with this, genistein was also recently found to prevent the proliferation, invasion, and migration and to induce apoptosis along with typical cell cycle arrest at the G2/M phase in breast cancer cells (Ye et al., 2018). These effects of genistein were caused by a decrease of S-phase kinase-associated protein 2 (Skp2), which is associated with the pathogenesis of multiple human cancers, along with an increase of the downstream tumor suppressors p21 and p27. These results indicate that the effects of genistein may be mediated through cell cycle regulation by changes of various cell cycle-related genes and proteins.

4.5.2. Daidzein

Daidzein induces Bcl-2/Bax ratio along with activation of caspase-9 and -7 in MCF-7 cells, leading to their subsequent death from apoptosis (Jin et al., 2009). This finding indicates that daidzein-induced apoptosis occurs through the mitochondrial caspase-dependent cell death pathway in MCF-7 cells (Park et al., 2013). In comparison, daidzein induced the expression of BRF2 in ER-positive breast cancer cells (Koo et al., 2015). Notably, BRF2 constitutes a transcription factor required for the synthesis of a small group of non-coding RNAs and its overexpression induces the transformation of human mammary epithelial cells. Therefore, these mechanisms may both contribute to the anticancer effects of daidzein.

Furthermore, at 50 μM, daidzein and equol prevented tumor cell invasion and metastasis via the effects of matrix metalloproteinase 2 (MMP-2) and tissue inhibitor of metalloproteinases 2 (TIMP-2) in human breast cancer MDA-MB-231 cells (Magee et al., 2013). In addition, equol inhibited cell growth in a time- and dose-dependent manner and enhanced radiation-induced apoptosis in human breast cancer cell lines MDA-MB-231 and T47D (Taghizadeh et al., 2013). These results suggest that equol in particular may be effective in enhancing cell death following irradiation.

4.6. Prostate Cancer

In the United States, prostate cancer constitutes one of the most common cancers in men (Pavese et al., 2014). Several epidemiological studies have indicated a relationship between phytoestrogens and prostate cancer risk (He et al., 2015; Zhang et al., 2017). However, another recent meta-analysis found no strong data to support that circulating levels of isoflavones are associated with prostate cancer risk (Perez-Cornago et al., 2018). Nevertheless, genistein and daidzein are believed to be strong anticancer agents and have anticancer properties (Adjakly et al., 2013).

4.6.1. Genistein

At high isoflavone concentration, genistein prevented cell detachment, protease production, cell invasion, and metastasis in human prostate cancer (Pavese et al., 2014). An important factor in inhibiting cancer progression is preventing the metastasis of the cancer. Therefore, genistein may inhibit cancer progression by inhibiting metastasis in human prostate tissue. Alternatively, genistein was shown to demonstrate protective effects against prostate cancer through the modulation of ERβ (Mahmoud et al., 2015), which plays a suppressive role to inhibit the formation and progression of prostate cancer.

The structural similarities between genistein and estrogen underlie its effects on the ERβ. Furthermore, it was shown that genistein inhibited the cell growth of prostate cancer through the regulation of miRNAs. miR-34a in prostate cancer cells (PC3 and DU145) (Chiyomaru et al., 2013a). In addition, genistein enhanced tumor suppressor miR-574-3p expression, which targets several cell signaling pathways such as the Jak-STAT and Wnt signaling pathways (Chiyomaru et al., 2013b). Enhanced miR-574-3p induces apoptosis by reducing Bcl-xL and activating caspase-9 and -3. These results suggest that genistein may inhibit prostate cancer through the regulation of various microRNAs.

4.6.2. Daidzein

A Japanese nested case-control study indicated that genistein levels tended to be inversely associated with the risk of prostate cancer (Kurahashi et al., 2008). As Japan has a lower incidence of prostate cancer than the West, the increased daidzein intake may also contribute to the inhibition of prostate cancer. In mouse, daidzein inhibited metastasis to the lymph nodes in PC-3 prostate tumors (Singh-Gupta et al., 2010). Daidzein also inhibited the effects of NF-κB and HIF-1alpha and subsequently inhibited the growth of the cancer cells such as PC-3 and C4-2B cells.

In comparison, equol inhibit cell growth and induced apoptosis via the Akt/FOXO3a pathway in human prostate cancer cell lines such as PC3, DU145, and LnCaP (Lu et al., 2016). In addition, *in vitro* investigation indicated that equol prevented the growth of PC3 xenograft tumors in BALB/c nude mice. Equol also exhibited significant anti-invasion effect in DU145 cells *in vitro* (Zheng et al., 2012). Specifically, equol inhibited cell invasion through the decrease of matrix metalloproteinase-9, -2, urokinase-type plasminogen activator, and SOD by antioxidant activity in prostate cancer DU145 cells. The effects of daidzein and its metabolite equol may therefore contribute to the anti-prostate cancer effects of soy.

4.7. Osteoporosis

Osteoporosis is characterized by low bone mass, which induces the deterioration of bone tissues and enhances the risk of fracture (Cummings and Melton, 2002). The prevalence of osteoporosis is 6–28% in men and women 50 years or older (Miao et al., 2012). In women in particular, the lack of postmenopausal estrogen plays a critical role for bone repair and maintenance (Nelson, 2008). Isoflavone may therefore influence osteoporosis via an estrogenic effect.

4.7.1. Genistein

In postmenopausal women, genistein was demonstrated to show anti-osteoporotic activity in post-hoc analysis of a multicenter randomized controlled trial (Arcoraci et al., 2017).

4.7.2. Daidzein

Daidzein induced cell proliferation by altering the cell cycle, enhanced cell differentiation by stimulating alkaline phosphatase activity and collagen levels, and decreased cell apoptosis in human osteoblast-like MG-63 cells (Jin et al., 2017). These effects of daidzein were shown to occur through the activation of MEK/ERK and PI3K/Akt in an ER-dependent manner. Furthermore, in ovariectomized rats, the combination of daidzein and gold kiwifruit-supplemented diet reduced the bone loss enhanced by estrogen deficiency (Tousen et al., 2013). Moreover, in estrogen-deficient osteoporotic mice, equol administration prevented bone loss and fat accumulation (Ishimi 2010).

4.8. Stroke

Stroke is major cause of worldwide fatality and disability, with approximately 50% of stroke survivors becoming dependent on caregivers for daily activities (Gillespie and Campbell, 2012). Stroke is considered to represent a cerebrovascular disorder, and can be classified as cerebral infarction, cerebral hemorrhage, and subarachnoid hemorrhage. Isoflavones such as genistein or daidzein have been shown to exhibit protective effects against stroke (Banecka-Majkutewicz et al., 2012).

4.8.1. Genistein

Genistein is considered likely to induce a neuroprotective effect in patients with stroke. For example, genistein reduced infarct volumes and decreased neuronal apoptosis at 24 h after reperfusion in the middle cerebral artery occlusion (MCAO) mouse (Wang et al., 2014). In addition, in ovariectomized MCAO rats, genistein decreased infarct volume and the production of reactive oxygen species (ROS), and enhanced the expression of Nrf2 and NQO1 but reduced that of cleaved-caspase 3 (Miao et al., 2018). These findings indicate that genistein prevents the oxidative stress injury induced by cerebral ischemia in ovariectomized rats by enhancing the Nrf2/NQO1 antioxidant system. Furthermore, genistein reduced I/R-induced thromboxane A2 and the increased leukocyte-platelet aggregates in rats (Cortina et al., 2013). These results further suggest that genistein limits the effect of circulatory function on neuronal dysfunction during I/R.

In addition, in rat cerebral ischemia, genistein reduced oxidative stress through the expression of SOD, and reduced neuronal degeneration by inhibiting apoptosis (Aras et al., 2015a). In H_2O_2-induced cell death in primary rat cortical neurons, genistein inhibited neuronal viability loss and prevented ROS generation and neuronal apoptosis (Qian et al., 2015). These effects of genistein were associated with the induction NF-κB/p65 and MAPK-signaling pathway regulation such as via JNK and ERK during H_2O_2 stimulation. Such genistein effects may be associated with inhibition of the strong ROS induction during hypoxia and reperfusion.

4.8.2. Daidzein

In the MCAO model, daidzein exerted neuroprotective effects during cerebral ischemia (Aras et al., 2015b). Similarly, daidzein reduced oxidative stress and neuronal apoptosis. These data suggest that daidzein exerts neuroprotective effects on ischemic brain tissue by inhibiting oxygen free radical production. Daidzein inhibited the neuronal death induced upon oxygen-glucose deprivation and simultaneously increased PPARγ activation (Hurtado et al., 2012), suggesting that daidzein inhibits neuronal death through activation of PPARγ. Notably, I/R induce neuronal injury in rats along with the expression of pp91(phox) and Src (Yu et al., 2014), which play a key role in mediating ischemic alteration in neurons. Conversely, equol decreased the mortality, infarct volume, serum lactate dehydrogenase activity, and malondialdehyde contents following I/R and reduced pp91(phox) and Src. These finding suggest that equol promoted neuroprotection in rats by inhibiting the activation of Src and upregulation of pp91(phox) induced by I/R.

CONCLUSION

Soybeans contain abundant protein, complex carbohydrates, dietary fiber, and low content of saturated fat. Soybeans also serve as a source of trace nutrients and isoflavones including daidzein and genistein. These isoflavones reduce the risk of dyslipidemia, ischemic heart disease, cancer, diabetes, and osteoporosis and relieve climacteric symptoms (Low Dog, 2005). Notably, the effects of these isoflavones suggest their mechanisms for moderating chronic disease risk. Therefore, the consumption of a sufficient amount of soybean foods might contribute to mild health benefits against lifestyle-related diseases such as those mentioned above. In this review, we described the protective effects of isoflavones against lifestyle-related diseases and their preventive mechanisms. Moreover, the metabolites of isoflavones also induce different effects. However, additional evidence of the beneficial role of isoflavone intake in these diseases is necessary, especially in human. Furthermore, the clinical benefit of these isoflavones needs to be confirmed.

REFERENCES

Adjakly, M., Ngollo, M., Boiteux, J. P., Bignon, Y. J, Guy, L. & Bernard-Gallon, D. (2013). Genistein and daidzein: different molecular effects on prostate cancer. *Anticancer Res*, *33*(1), 39-44.

Akiyama, T., Ishida, J., Nakagawa, S., Ogawara, H., Watanabe, S., Itoh, N., Shibuya, M. & Fukami, Y. (1987). Genistein, a specific inhibitor of tyrosine-specific protein kinases. *J Biol Chem*, *262*(12), 5592-5595.

Ambra, R., Rimbach, G., de Pascual, Teresa, S., Fuchs, D., Wenzel, U., Daniel, H. & Virgili, F. (2006). Genistein affects the expression of genes involved in blood pressure regulation and angiogenesis in primary human endothelial cells. *Nutr Metab Cardiovasc Dis*, *16*(1), 35-43.

Andres, S., Abraham, K., Appel, K. E. & Lampen, A. (2011). Risks and benefits of dietary isoflavones for cancer. *Crit Rev Toxicol*, *41*(6), 463-506.

Anthony, M. S., Clarkson, T. B., Bullock, B. C. & Wagner, J. D. (1997). Soy protein versus soy phytoestrogens in the prevention of diet-induced coronary artery atherosclerosis of male cynomolgus monkeys. *Arterioscler Thromb Vasc Biol*, *17* (11), 2524-2531.

Applegate, C. C., Rowles, J. L., Ranard, K. M., Jeon, S. & Erdman, J. W. (2018). Soy consumption and the risk of prostate cancer: an updated systematic review and meta-analysis. *Nutrients*, *10*(1). pii. E40.

Aras, A. B., Guven, M., Akman, T., Alacam, H., Kalkan, Y., Silan, C. & Cosar, M. (2015a). Genistein exerts neuroprotective effect on focal cerebral ischemia injury in rats. *Inflammation* . *38*(3), 1311-1321.

Aras, A. B., Guven, M., Akman, T., Ozkan, A., Sen, H. M., Duz, U., Kalkan, Y., Silan, C. & Cosar, M. (2015b). Neuroprotective effects of daidzein on focal cerebral ischemia injury in rats. *Neural Regen Res.*, Jan, *10*(1), 146-52. doi:10.4103/1673-5374.150724.

Arcoraci, V., Atteritano, M., Squadrito, F., D'Anna, R., Marini, H., Santoro, D., Minutoli. L., Messina, S., Altavilla, D. & Bitto. A. (2017). Antiosteoporotic activity of genistein aglycone in postmenopausal women: evidence from a post-hoc analysis of a multicenter randomized controlled trial. *Nutrients*, *9*(2). pii. E179.

Babu, P. V., Si, H., Fu, Z., Zhen, W. & Liu, D. (2012). Genistein prevents hyperglycemia-induced monocyte adhesion to human aortic endothelial cells through preservation of the cAMP signaling pathway and ameliorates vascular inflammation in obese diabetic mice. *J Nutr*, *142*(4), 724-730.

Banecka-Majkutewicz, Z., Sawula, W., Kadzinski, L., Wegrzyn, A. & Banecki, B. Homocysteine, heat shock proteins, genistein and vitamins in ischemic stroke--pathogenic and therapeutic implications. *Acta Biochim Pol*, *59*(4), 495-499.

Birru, R. L., Ahuja, V., Vishnu, A., Evans, R. W., Miyamoto, Y., Miura, K., Usui, T. & Sekikawa, A. (2016). The impact of equol-producing status in modifying the effect of soya isoflavones on risk factors for CHD: a systematic review of randomised controlled trials. *J Nutr Sci*, *5*, e30.

Cao, Y. X., Yang, X. J., Liu, J. & Li, K. X. (2006). Effects of daidzein sulfates on blood pressure and artery of rats. *Basic Clin Pharmacol Toxicol*, *99*(6), 425-430.

Chen, J, Duan, Y., Zhang, X., Ye, Y., Ge, B. & Chen, J. (2015). Genistein induces apoptosis by the inactivation of the IGF-1R/p-Akt signaling pathway in MCF-7 human breast cancer cells. *Food Funct*, *6*(3), 995-1000.

Cheong, S. H., Furuhashi, K., Ito, K., Nagaoka, M., Yonezawa, T., Miura, Y. & Yagasaki, K. (2014). Daidzein promotes glucose uptake through glucose transporter 4 translocation to plasma membrane in L6 myocytes and improves glucose homeostasis in Type 2 diabetic model mice. *J Nutr Biochem*, *25*(2), 136-143.

Chiyomaru, T., Yamamura, S., Fukuhara, S., Yoshino, H., Kinoshita, T., Majid, S., Saini, S., Chang, I., Tanaka, Y., Enokida, H., Seki, N., Nakagawa, M. & Dahiya, R. (2013a). Genistein inhibits prostate cancer cell growth by targeting miR-34a and oncogenic HOTAIR. *PLoS One*, *8*(8), e70372.

Chiyomaru, T., Yamamura, S., Fukuhara, S., Hidaka, H., Majid, S., Saini, S., Arora, S., Deng, G., Shahryari, V., Chang, I., Tanaka, Y., Tabatabai, Z. L., Enokida, H., Seki, N., Nakagawa, M. & Dahiya, R. (2013b). Genistein up-regulates tumor suppressor microRNA-574-3p in prostate cancer. *PLoS One*, *8*(3), e58929.

Cho, T. M., Peng, N., Clark, J. T., Novak, L., Roysommuti, S., Prasain, J. & Wyss, J. M. (2007). Genistein attenuates the hypertensive effects of dietary NaCl in hypertensive male rats. *Endocrinology*, *148*(11), 5396-5402.

Choi, E. J. & Kim, G. H. (2013). Antiproliferative activity of daidzein and genistein may be related to ERα/c-erbB-2 expression in human breast cancer cells. *Mol Med Rep*, *7*(3), 781-784.

Corbett, A. H., Hong, D. & Osheroff, N. (1993). Exploiting mechanistic differences between drug classes to define functional drug interaction domains on topoisomerase II. Evidence that several diverse DNA cleavage-enhancing agents share a common site of action on the enzyme. *J Biol Chem*, *268*(19), 14394-14398.

Cortina, B., Torregrosa, G., Castello-Ruiz, M., Burguete, M. C., Moscardo, A., Latorre, A., Salom, J. B., Valles, J., Santos, M. T. & Alborch, E. (2013). Improvement of the circulatory function partially accounts for the neuroprotective action of the phytoestrogen genistein in experimental ischemic stroke. *Eur J Pharmacol*, *708*(1-3), 88-94.

Cummings, S. R. & Melton, L. J. (2001). Epidemiology and outcomes of osteoporotic fractures. *Lancet*, *359*(9319), 1761-1767.

Das, D., Sarkar, S., Bordoloi, J., Wann, S. B., Kalita, J. & Manna, P. (2018). Daidzein, its effects on impaired glucose and lipid metabolism and vascular inflammation associated with type 2 diabetes. *Biofactors*, *44*(5), 407-417.

de Kleijn, M. J., van der Schouw, Y. T., Wilson, P. W., Grobbee, D. E. & Jacques P. F. (2002). Dietary intake of phytoestrogens is associated with a favorable metabolic cardiovascular risk profile in postmenopausal U.S. women: the Framingham study. *J Nutr*, *132*(2), 276-282.

Ding, M., Franke, A. A., Rosner, B. A, Giovannucci, E., van Dam R. M., Tworoger, S. S., Hu, F. B. & Sun, Q. (2015). Urinary isoflavonoids and risk of type 2 diabetes: a prospective investigation in US women. *Br J Nutr*, *114*(10), 1694-1701.

El-Kordy, E. A. & Alshahrani, A. M. (2015). Effect of genistein, a natural soy isoflavone, on pancreatic β-cells of streptozotocin-induced diabetic rats: Histological and immunohistochemical study. *J Microsc Ultrastruct*, *3*(3), 108-119.

Erdman, J. W. Jr., Badger, T. M., Lampe, J. W., Setchell, K. D. & Messina, M. (2004). Not all soy products are created equal: caution needed in interpretation of research results. *J Nutr*, *134*(5), 1229S-1233S.

Ferlay, J., Soerjomataram, I., Dikshit, R., Eser, S., Mathers, C., Rebelo, M., Parkin, D. M., Forman, D. & Bray, F. (2015). Cancer incidence and mortality worldwide: sources, methods and major patterns in GLOBOCAN 2012. *Int J Cancer*, *136*(5), E359-E386.

Gao, Y., Yao, Y., Zhu, Y. & Ren, G. (2015). Isoflavones in chickpeas inhibit adipocyte differentiation and prevent insulin resistance in 3T3-L1 cells. *J Agric Food Chem*, *63*(44), 9696-9703.

Gil-Izquierdo, A., Penalvo, J. L., Gil, J. I., Medina, S., Horcajada, M. N., Lafay, S., Silberberg, M., Llorach, R., Zafrilla, P., Garcia-Mora, P. & Ferreres, F. (2012). Soy isoflavones and cardiovascular disease epidemiological, clinical and -omics perspectives. *Curr Pharm Biotechnol*, *13*(5), 624-631.

Gillespie, D. & Campbell, F. (2011). Effect of stroke on family carers and family relationships. *Nurs Stand*, *26*(2), 39-46.

Gupta, S. K., Dongare, S., Mathur, R., Mohanty, I. R., Srivastava, S., Mathur, S. & Nag, T. C. (2015). Genistein ameliorates cardiac inflammation and oxidative stress in streptozotocin-induced diabetic cardiomyopathy in rats. *Mol Cell Biochem*, *408*(1-2), 63-72.

Hall, W. L., Vafeiadou, K., Hallund, J., Bugel, S., Reimann, M., Koebnick, C., Zunft, H. J., Ferrari, M., Branca, F., Dadd, T., Talbot, D., Powell, J., Minihane, A. M., Cassidy, A., Nilsson, M., Dahlman-Wright, K., Gustafsson, J. A. & Williams, C. M. (2006). Soy-isoflavone-enriched foods and markers of lipid and glucose metabolism in postmenopausal women: interactions with genotype and equol production. *Am J Clin Nutr*, *83*(3), 592-600.

Haneishi, A., Takagi, K., Asano, K., Nakamura, S., Kagawa, N. & Yamada, K. (2011). Genistein stimulates the insulin-dependent signaling pathway. *Front Biosci (Elite Ed).*, *3*, 1534-1540.

Haskell, W. L. (2003). Cardiovascular disease prevention and lifestyle interventions, effectiveness and efficacy. *J Cardiovasc Nurs*, *18*(4), 245-255.

He, J., Wang, S., Zhou, M., Yu, W., Zhang, Y. & He, X. (2015). Phytoestrogens and risk of prostate cancer: a meta-analysis of observational studies. *World J Surg Oncol*, *13*, 231.

Hintz, K. K. & Ren, J. (2004). Phytoestrogenic isoflavones daidzein and genistein reduce glucose-toxicity-induced cardiac contractile dysfunction in ventricular myocytes. *Endocr Res*, *30*(2), 215-223.

Horiuchi, H., Usami, A., Shirai, R., Harada, N., Ikushiro, S., Sakaki, T., Nakano, Y., Inui, H. & Yamaji, R. (2017). S-Equol activates cAMP signaling at the plasma membrane of INS-1 pancreatic β-cells and protects against streptozotocin-induced hyperglycemia by increasing β-cell function in male mice. *J Nutr*, *147*(9), 1631-1639.

Hu, W. S., Lin, Y. M., Ho, T. J., Chen, R. J., Li, Y. H., Tsai, F. J., Tsai, C. H., Day, C. H., Chen, T. S. & Huang, C. Y. (2013). Genistein suppresses the isoproterenol-treated H9c2 cardiomyoblast cell apoptosis associated with P-38, Erk1/2, JNK, and NFκB signaling protein activation. *Am J Chin Med*, *41*(5), 1125-1136.

Hu, W. S., Lin, Y. M., Kuo, W. W., Pan, L. F., Yeh, Y. L., Li, Y. H., Kuo, C. H., Chen, R. J., Padma, V. V., Chen, T. S. & Huang, C. Y. (2016). Suppression of isoproterenol-induced apoptosis in H9c2 cardiomyoblast cells by daidzein through activation of Akt. *Chin J Physiol*, *59*(6), 323-330.

Hurtado, O., Ballesteros, I., Cuartero, M. I., Moraga, A., Pradillo, J. M., Ramirez-Franco, J., Bartolome-Martin, D., Pascual, D., Torres, M., Sanchez-Prieto, J., Salom, J. B., Lizasoain, I. & Moro, M. A. (2012). Daidzein has neuroprotective effects through ligand-binding-independent PPARγ activation. *Neurochem Int*, *61*(1), 119-127.

Ishimi, Y. (2010). Dietary equol and bone metabolism in postmenopausal Japanese women and osteoporotic mice. *J Nutr*, *140*(7), 1373S-1376S.

Ji, E. S., Yue, H., Wu, Y. M. & He, R. R. (2004). Effects of phytoestrogen genistein on myocardial ischemia/reperfusion injury and apoptosis in rabbits. *Acta Pharmacol Sin*, *25*(3), 306-312.

Jin, L., Zhao, X., Qin, Y., Zhu, W., Zhang, W., Liu, A. & Luo, Z. (2015). Soy isoflavones protect against H2O2-induced injury in human umbilical vein endothelial cells. *Mol Med Rep*, *12*(3), 4476-4482.

Jin, X., Sun, J., Yu, B., Wang, Y., Sun, W. J., Yang, J., Huang, S. H. & Xie, W. L. (2017). Daidzein stimulates osteogenesis facilitating proliferation, differentiation, and antiapoptosis in human osteoblast-like MG-63 cells via estrogen receptor-dependent MEK/ERK and PI3K/Akt activation. *Nutr Res*, *42*, 20-30.

Joannou, G. E., Kelly, G. E., Reeder, A. Y., Waring, M. & Nelson, C. (1995). A urinary profile study of dietary phytoestrogens. The identification and mode of metabolism of new isoflavonoids. *J Steroid Biochem Mol Biol*, *54*, 167-184.

Kim, Y. S., Kim, N. H., Jung, D. H., Jang, D. S., Lee, Y. M., Kim, J. M. & Kim, J. S. (2008). Genistein inhibits aldose reductase activity and high glucose-induced TGF-beta2 expression in human lens epithelial cells. *Eur J Pharmacol.*, *594*(1-3), 18-25.

Kim, J. W., Jin, Y. C., Kim, Y. M., Rhie, S., Kim, H. J., Seo, H. G., Lee, J. H., Ha, Y. L., Chang, K. C. (2009). Daidzein administration *in vivo* reduces myocardial injury in a rat ischemia/reperfusion model by inhibiting NF-kappaB activation. *Life Sci*, *84*(7-8), 227-234.

Kim, M. J. & Lim, Y. (2013). Protective effect of short-term genistein supplementation on the early stage in diabetes-induced renal damage. *Mediators Inflamm. 2013*, 2013, 510212.

Ko, K. P. (2014). Isoflavones: chemistry, analysis, functions and effects on health and cancer. *Asian Pac J Cancer Prev*, *15*(17), 7001-7110.

Kobayashi, M., Egusa, S. & Fukuda, M. (2014). Isoflavone and protein constituents of lactic acid-fermented soy milk combine to prevent dyslipidemia in rats fed a high cholesterol diet. *Nutrients*, *6*(12), 5704-5723.

Koo, J., Cabarcas-Petroski, S., Petrie, J. L., Diette, N., White, R. J. & Schramm, L. (2015). Induction of proto-oncogene BRF2 in breast cancer cells by the dietary soybean isoflavone daidzein. *BMC Cancer*, *15*, 905.

Kurahashi, N., Iwasaki, M., Inoue, M., Sasazuki, S. & Tsugane, S. (2008). Plasma isoflavones and subsequent risk of prostate cancer in a nested case-control study: the Japan Public Health Center. *J Clin Oncol*, *26*(36), 5923-5929.

Lampe, J. W. (2003). Isoflavonoid and lignan phytoestrogens as dietary biomarkers. *J Nutr*, *133* Suppl 3, 956S-9964S.

Lee, J. S. (2006). Effects of soy protein and genistein on blood glucose, antioxidant enzyme activities, and lipid profile in streptozotocin-induced diabetic rats. *Life Sci*, *79*(16), 1578-1584.

Li, H. F., Wang, L. D. & Qu, S. Y. (2004). Phytoestrogen genistein decreases contractile response of aortic artery *in vitro* and arterial blood pressure *in vivo*. *Acta Pharmacol Sin*, *25*(3), 313-318.

Li, W., Ruan, W., Peng, Y. & Wang, D. (2018). Soy and the risk of type 2 diabetes mellitus: A systematic review and meta-analysis of observational studies. *Diabetes Res Clin Pract*, *137*, 190-199.

Li, Y. & Zhang, H. (2017). Soybean isoflavones ameliorate ischemic cardiomyopathy by activating Nrf2-mediated antioxidant responses. *Food Funct.*, *8*(8), 2935-2944.

Low Dog, T. (2005). Menopause: a review of botanical dietary supplements. *Am J Med*, *118* Suppl 12B, 98-108

Lu, Z., Zhou, R., Kong, Y., Wang, J., Xia, W., Guo, J., Liu, J., Sun, H., Liu, K., Yang, J., Mi, M. & Xu, H. (2016). S-equol, a secondary metabolite of natural anticancer isoflavone daidzein, inhibits prostate cancer growth *in vitro* and *in vivo*, though activating the Akt/FOXO3a pathway. *Curr Cancer Drug Targets*, *16*(5), 455-465.

Mahn, K., Borras, C., Knock, G. A., Taylor, P., Khan, I. Y., Sugden, D., Poston, L., Ward, J. P., Sharpe, R. M., Vina, J., Aaronson, P. I. & Mann, G. E. (2005). Dietary soy isoflavone induced increases in antioxidant and eNOS gene expression lead to improved endothelial function and reduced blood pressure *in vivo*. *FASEB J*, *19*(12), 1755-1757.

Mahmoud, A. M., Al-Alem, U., Ali, M. M. & Bosland, M. C. (2015). Genistein increases estrogen receptor beta expression in prostate cancer via reducing its promoter methylation. *J Steroid Biochem Mol Biol*, *152*, 62-75.

Magee, P. J., Allsopp, P., Samaletdin, A. & Rowland, I. R. (2014). Daidzein, R-(+) equol and S-(-) equol inhibit the invasion of MDA-MB-231 breast cancer cells potentially via the down-regulation of matrix metalloproteinase-2. *Eur J Nutr*, *53*(1), 345-350.

Mateos-Aparicio, I., Redondo Cuenca, A., Villanueva-Suarez, M. J. & Zapata-Revilla, M. A. (2008). Soybean, a promising health source. *Nutr Hosp*, *23*(4), 305-312.

Maulik, S. K., Prabhakar, P., Dinda, A. K. & Seth, S. (2012). Genistein prevents isoproterenol-induced cardiac hypertrophy in rats. *Can J Physiol Pharmacol.*, *90*(8), 1117-1125.

Messina, M. J. (1999). Legumes and soybeans: overview of their nutritional profiles and health effects. *Am J Clin Nutr.*, *70*(3 Suppl), 439S-450S.

Miao, Z. Y., Xia, X., Che, L. & Song, Y. T. (2018). Genistein attenuates brain damage induced by transient cerebral ischemia through up-regulation of Nrf2 expression in ovariectomized rats. *Neurol Res*, *40*(8), 689-695.

Miao, Q., Li, J. G., Miao, S., Hu, N., Zhang, J., Zhang, S., Xie, Y. H., Wang, J. B. & Wang, S. W. (2012). The bone-protective effect of genistein in the animal model of bilateral ovariectomy: roles of phytoestrogens and PTH/PTHR1 against post-menopausal osteoporosis. *Int J Mol Sci*, *13*(1), 56-70.

Montenegro, M. F., Pessa, L. R. & Tanus-Santos, J. E. (2009). Isoflavone genistein inhibits the angiotensin-converting enzyme and alters the vascular responses to angiotensin I and bradykinin. *Eur J Pharmacol*, *607*(1-3), 173-177

Nadal-Serrano, M., Pons, D. G., Sastre-Serra, J., Blanquer-Rossello, Mdel, M., Roca, P. & Oliver, J. (2013). Genistein modulates oxidative stress in breast cancer cell lines according to ERα/ERβ ratio: effects on mitochondrial functionality, sirtuins, uncoupling protein 2 and antioxidant enzymes. *Int J Biochem Cell Biol*, *45*(9), 2045-2051.

Nestel, P., Fujii, A. & Zhang, L. (2007). An isoflavone metabolite reduces arterial stiffness and blood pressure in overweight men and postmenopausal women. *Atherosclerosis*, *192*(1), 184-189.

Nelson, H. D. (2008). Menopause. *Lancet*, *371*(9614), 760-70.

Omoni, A. O. & Aluko, R. E. (2005). Soybean foods and their benefits: potential mechanisms of action. *Nutr Rev*, *63*(8), 272-283.

Palanisamy, N. & Venkataraman, A. C. (2013). Beneficial effect of genistein on lowering blood pressure and kidney toxicity in fructose-fed hypertensive rats. *Br J Nutr*, *109*(10), 1806-1812.

Paterni, I., Granchi, C., Katzenellenbogen, J. A. & Minutolo, F. (2014). Estrogen receptors alpha (ERα) and beta (ERβ): subtype-selective ligands and clinical potential. *Steroids*, *90*, 13-29.

Park, H. J., Jeon, Y. K., You, D. H. & Nam, M. J. (2013). Daidzein causes cytochrome c-mediated apoptosis via the Bcl-2 family in human hepatic cancercells. *Food Chem Toxicol*, *60*, 542-549.

Pavese, J. M., Krishna, S. N. & Bergan, R. C. (2014). Genistein inhibits human prostate cancer cell detachment, invasion, and metastasis. *Am J Clin Nutr*, *100* Suppl 1, 431S-436S.

Pepine, C. J., von Mering, G. O., Kerensky, R. A., Johnson, B. D., McGorray, S. P., Kelsey, S. F., Pohost, G., Rogers, W. J., Reis, S. E., Sopko, G., Bairey Merz, C. N. & WISE Study Group. (2007). Phytoestrogens and coronary microvascular function in women with suspected myocardial ischemia: a report from the Women's Ischemia Syndrome Evaluation (WISE) Study. *J Womens Health (Larchmt)*, *16*(4), 481-488.

Perez-Cornago, A., Appleby, P. N., Boeing, H., Gil, L., Kyrphi, C., Ricceri, F., Murphy, N., Trichopoulou, A., Tsilidis, K. K., Khaw, K. T., Luben, R. N., Gislefoss, R. E., Langseth, H., Drake, I., Sonestedt, E., Wallstrom, P., Stattin, P., Johansson, A., Landberg, R., Nilsson, L. M., Ozasa, K., Tamakoshi, A., Mikami, K., Kubo, T., Sawada, N., Tsugane, S., Key, T. J., Allen, N. E. & Travis, R. C. (2018). Circulating isoflavone and lignan concentrations and prostate cancer risk: a meta-analysis of individual participant data from seven prospective studies including 2,828 cases and 5,593 controls. *Int J Cancer*, *43*(11), 2677-2686.

Qian, Y., Cao, L., Guan, T., Chen, L., Xin, H., Li, Y., Zheng, R. & Yu, D. (2015). Protection by genistein on cortical neurons against oxidative stress injury via inhibition of NF-kappaB, JNK and ERK signaling pathway. *Pharm Biol*, *53*(8), 1124-1132.

Rivas, M., Garay, R. P., Escanero, J. F., Cia, P., Jr., Cia, P. & Alda, J. O. (2002). Soy milk lowers blood pressure in men and women with mild to moderate essential hypertension. *J Nutr*, *132*(7), 1900-1902.

Redondo-Cuenca, A., Villanueva-Suarez, M. J. & Mateos-Aparicio, I. (2008). Soybean seeds and its by-product okara as sources of dietary fibre. Measurement by AOAC and Englyst methods. *Food Chem*, *108*(3), 1099-1105.

Rigalli, J. P., Tocchetti, G. N., Arana, M. R., Villanueva, S. S., Catania, V. A., Theile, D., Ruiz, M. L. & Weiss, J. (2016). The phytoestrogen genistein enhances multidrug resistance in breast cancer cell lines by translational regulation of ABC transporters. *Cancer Lett*, *376*(1), 165-72.

Romagnolo, D. F., Donovan, M. G., Papoutsis, A. J., Doetschman, T. C. & Selmin, O. I. (2017). Genistein prevents BRCA1 CpG methylation and proliferation in human breast cancer cells with activated aromatic hydrocarbon receptor. *Curr Dev Nutr*, *1*(6), e000562.

Prietsch, R. F., Monte, L. G., da Silva, F. A., Beira, F. T., Del Pino, F. A., Campos, V. F., Collares, T., Pinto, L. S., Spanevello, R. M., Gamaro, G. D. & Braganhol, E. (2014). Genistein induces apoptosis and autophagy in human breast MCF-7 cells by modulating the expression of proapoptotic factors and oxidative stress enzymes. *Mol Cell Biochem*, *390*(1-2), 235-242.

Ryu, Y. G., Won, B., Park, H. R., Ghafoor, K. & Park, J. (2010). Effects of the beta-glycosidase reaction on bio-conversion of isoflavones and quality during tofu processing. *J Sci Food Agric*, *90*(5), 843-849.

Schabath, M. B., Hernandez, L. M., Wu, X., Pillow, P. C. & Spitz, M. R. (2005). Dietary phytoestrogens and lung cancer risk. *JAMA*, *294*(12), 1493-1504.

Si, H. & Liu, D. (2008). Genistein, a soy phytoestrogen, upregulates the expression of human endothelial nitric oxide synthase and lowers blood pressure in spontaneously hypertensive rats. *J Nutr*, *138*(2), 297-304.

Singh-Gupta, V., Zhang, H., Yunker, C. K., Ahmad, Z., Zwier, D., Sarkar, F. H. & Hillman, G. G. (2010). Daidzein effect on hormone refractory prostate cancer *in vitro* and *in vivo* compared to genistein and soy extract: potentiation of radiotherapy. *Pharm Res*, *27*(6), 1115-27.

Taghizadeh, B., Ghavami, L., Nikoofar, A. & Goliaei, B. (2015). Equol as a potent radiosensitizer in estrogen receptor-positive and -negative human breast cancer cell lines. *Breast Cancer.*, *22*(4), 382-390.

Tang, C., Zhang, K., Zhao, Q. & Zhang, J. (2015). Effects of Dietary Genistein on Plasma and Liver Lipids, Hepatic Gene Expression, and Plasma Metabolic Profiles of Hamsters with Diet-Induced Hyperlipidemia. *J Agric Food Chem*, *63*(36), 7929-7936.

Tang, Y., Li, S., Zhang, P., Zhu, J., Meng, G., Xie, L., Yu, Y., Ji, Y. & Han, Y. (2016). Soy Isoflavone Protects Myocardial Ischemia/ Reperfusion Injury through Increasing Endothelial Nitric Oxide Synthase and Decreasing Oxidative Stress in Ovariectomized Rats. *Oxid Med Cell Longev*, 5057405. doi: 10.1155/2016/5057405.

Toda, T., Uesugi, T., Hirai, K., Nukaya, H., Tsuji, K. & Ishida, H. (1999). New 6-O-acyl isoflavone glycosides from soybeans fermented with Bacillus subtilis (natto). I. 6-O-succinylated isoflavone glycosides and their preventive effects on bone loss in ovariectomized rats fed a calcium-deficient diet. *Biol Pharm Bull*, *22*(11), 1193-1201.

Tormala, R. M., Appt, S., Clarkson, T. B., Tikkanen, M. J., Ylikorkala, O. & Mikkola, T. S. (2007). Individual differences in equol production capability modulate blood pressure in tibolone-treated postmenopausal women: lack of effect of soy supplementation. *Climacteric*, *10*(6), 471-479.

Tousen, Y., Wolber, F. M, Chua, W. H., Tadaishi, M., Ishimi, Y. & Kruger, M. C. (2014). Effects of daidzein and kiwifruit on bone mineral density and equol production in ovariectomised rats. *Int J Food Sci Nutr*, *65*(3), 360-377.

Tsai, C. Y., Chen, Y. H., Chien, Y. W., Huang, W. H. & Lin, S. H. (2010). Effect of soy saponin on the growth of human colon cancer cells. *World J Gastroenterol*, *16*(27), 3371-3376.

Tse, G. & Eslick, G. D. (2016). Soy and isoflavone consumption and risk of gastrointestinal cancer: a systematic review and meta-analysis. *Eur J Nutr*, *55*(1), 63-73.

Tseng, L. M., Liu, C. Y., Chang, K. C., Chu, P. Y., Shiau, C. W. & Chen, K. F. (2012). CIP2A is a target of bortezomib in human triple negative breast cancer cells. *Breast Cancer Res*, *14*(2), R68.

Uifalean, A., Schneider, S., Ionescu, C., Lalk, M. & Iuga, C. A. (2015). Soy Isoflavones and Breast Cancer Cell Lines: Molecular Mechanisms and Future Perspectives. *Molecules*, *21*(1), E13.

Vitale, D. C., Piazza, C., Melilli, B., Drago, F. & Salomone, S. (2013). Isoflavones: estrogenic activity, biological effect and bioavailability. *Eur J Drug Metab Pharmacokinet.*, *38*(1), 15-25.

Wada, K., Nakamura, K., Tamai, Y., Tsuji, M., Kawachi, T., Hori, A., Takeyama, N., Tanabashi, S., Matsushita, S., Tokimitsu, N. & Nagata, C. (2013). Soy isoflavone intake and breast cancer risk in Japan: from the Takayama study. *Int J Cancer*, *133*(4), 952-960.

Wang H. J. & Murphy, P. A. (1994). Isoflavone content in commercial soybean foods. *J Agric Food Chem*, *42*, 1666–1673.

Wang, S., Wei, H., Cai, M., Lu, Y., Hou, W., Yang, Q., Dong, H. & Xiong, L. (2014). Genistein attenuates brain damage induced by transient cerebral ischemia through up-regulation of ERK activity in ovariectomized mice. *Int J Biol Sci*, *10*(4), 457-65.

Wang, T. T., Sathyamoorthy, N. & Phang, J. M. (1996). Molecular effects of genistein on estrogen receptor mediated pathways. *Carcinogenesis.*, *17*(2), 271-255.

Wong, J. M., Kendall, C. W., Marchie, A., Liu, Z., Vidgen, E., Holmes, C., Jackson, C. J., Josse, R. G., Pencharz, P. B., Rao, A. V., Vuksan, V., Singer, W. & Jenkins, D. J. (2012). Equol status and blood lipid profile in hyperlipidemia after consumption of diets containing soy foods. *Am J Clin Nutr.*, *95*(3), 564-571.

Yang, R., Jia, Q., Liu, X. F. & Ma, S. F. (2018). Effect of genistein on myocardial fibrosis in diabetic rats and its mechanism. *Mol Med Rep.*, *17*(2), 2929-2936.

Ye, D., Li, Z. & Wei, C. (2018). Genistein inhibits the S-phase kinase-associated protein 2 expression in breast cancer cells. *Exp Ther Med*, *15*(1), 1069-1075.

Yu, W., Wang, Y., Zhou, D. X., Zhao, L. M., Li, G. R. & Deng, X. L. (2014). Equol is neuroprotective during focal cerebral ischemia and reperfusion that involves p-Src and gp91(phox). *Curr Neurovasc Res*, *11*(4), 367-377.

Zung, A., Shachar, S., Zadik, Z. & Kerem, Z. (2010). Soy-derived isoflavones treatment in children with hypercholesterolemia: a pilot study. *J Pediatr Endocrinol Metab*, *23*(1-2), 133-141.

Zhang, L., Yang, B., Zhou, K., Li, H., Li, D., Gao, H., Zhang, T., Wei, D., Li, Z. & Diao, Y. (2015). Potential therapeutic mechanism of genistein in breast cancer involves inhibition of cell cycle regulation. *Mol Med Rep*, *11*(3), 1820-1826.

Zhang, M., Ikeda, K., Xu, J. W., Yamori, Y., Gao, X. M. & Zhang, B. L. (2009). Genistein suppresses adipogenesis of 3T3-L1 cells via multiple signal pathways. *Phytother Res*, *23*(5), 713-738.

Zhang, Q., Feng, H., Qluwakemi, B., Wang, J., Yao, S., Cheng, G., Xu, H., Qiu, H., Zhu, L. & Yuan, M. Phytoestrogens and risk of prostate cancer: an updated meta-analysis of epidemiologic studies. *Int J Food Sci Nutr*, *68*(1), 28-42.

Zhao, Q., Zhao, M., Parris, A. B., Xing, Y. & Yang, X. (2016). Genistein targets the cancerous inhibitor of PP2A to induce growth inhibition and apoptosis in breast cancer cells. *Int J Oncol*, *49*(3), 1203-1210.

Zheng, W., Zhang, Y., Ma, D., Shi, Y., Liu, C. & Wang, P. (2012). (+/-) Equol inhibits invasion in prostate cancer DU145 cells possibly via down-regulation of matrix metalloproteinase-9, matrix metalloproteinase-2 and urokinase-type plasminogen activator by antioxidant activity. *J Clin Biochem Nutr*, *51*(1), 61-67.

Zhong, W. W., Liu, Y. & Li, C. L. (2011). Mechanisms of genistein protection on pancreas cell damage in high glucose condition. *Intern Med*, *50*(19), 2129-2134.

In: Phytochemicals
Editor: Iman Ryan
ISBN: 978-1-53615-478-8

Chapter 3

THE EFFECT OF CRUDE EXTRACTS OF THE NUTRACEUTICAL FRUITS OF *VITEX PAYOS* ON HYPERGLYCAEMIA IN ALLOXAN-INDUCED DIABETIC RATS

***Lynette Maraura*[1], *Tafadzwa Taderera*[2], *Loveness Kuziwa Nyanga*[3] *and Cathrine Chidewe*[1,*]**
[1]Department of Biochemistry, Faculty of Science, University of Zimbabwe, Mount Pleasant, Harare, Zimbabwe
[2]Department of Physiology, College of Health Sciences, University of Zimbabwe, Mount Pleasant, Harare, Zimbabwe
[3]Institute of Food, Nutrition and Family Sciences, Faculty of Science, University of Zimbabwe, Mount Pleasant, Harare, Zimbabwe

ABSTRACT

This study sought to scientifically validate the hypoglycaemic effects of crude aqueous extracts from *Vitex payos* fruits on both normal and

* Corresponding Author's E-mail: cchidewe@gmail.com.

alloxan-induced diabetic rats. Powdered pulp of *V. payos* fruit was hydrated with distilled water to 0.27 g/ml and water soluble compounds were extracted. Alloxan-induced diabetic rats were used to determine the hypoglycaemic effect of the extract after 16 days of treatment in comparison to distilled water and glibenclamide controls. Glucose tolerance test was also done to determine the effectiveness of the plant extracts as a hypoglycaemic agent. Alloxan-induced diabetic rats in the negative control group given distilled water died before the second measurement of blood glucose on the third day of treatment. Over the 16 day period, a general decrease in blood glucose levels to 5.2 ± 0.3 mmol/L, was observed in rats that were treated with glibenclamide. In the alloxan-induced diabetic groups that were administered with crude aqueous extracts of *V. payos*, a dose dependent decrease in blood glucose levels was observed. The rats that were administered with 800mg/kg of extract had their glucose levels decreasing more than those that were administered with doses of 400mg/kg and 100mg/kg. There was no significant difference ($P > 0.05$) in blood glucose levels of normal rats that were administered with different concentrations of the crude extracts of *V. payos* fruits compared to the normal rats that were given distilled water (control group) over the 16 days of treatment. After initial glucose feeding in the OGTT test, an increase in blood glucose levels was observed in both normal and alloxan-induced diabetic rats that were administered with extracts of *V. payos*. After 60 minutes all the groups administered with extracts of *V. payos* had glucose levels gradually decreasing. In the OGTT test normal rats that were given distilled water (negative control group) had their glucose levels increasing until the end of 120 minutes. Rats that were treated with glibenclamide had an initial increase in glucose level over the first 30 minutes of glucose feeding after which glucose levels began to decrease. Crude aqueous extracts from fruits of *V. payos* showed effective hypoglycaemic activity in the alloxan-induced diabetic rats and could potentially work as a prophylactic in deterring hyperglycaemia.

Keywords: diabetes mellitus, *Vitex payos*, antidiabetic potential, alloxan-induced

1. Introduction

The word diabetes was first recorded in 1425, the Greek word mellitus was then added which means “like honey,” to illustrate the nature of the urine which was described as sweet for both taste and smell. Diabetes mellitus (DM) is a metabolic disorder of multiple physiology characterised by chronic hyperglycemia with disturbances of carbohydrate, fat and protein metabolism that result from defects in the secretion of insulin, action of the hormone, or both (Evcimen and King 2007; Safamansouri et al. 2014). DM can either develop in an individual or the tendency to develop the disease may be acquired genetically. When levels of insulin in the blood are low tissue fails to effectively absorb and utilise glucose available in the circulating blood. Hyperglycaemia or raised blood sugar is a common effect of uncontrolled DM and over time leads to serious damage to the nerves and blood vessels (Evcimen and King 2007).

The prevalence of DM has increased dramatically in middle and low income countries (Mathers and Lonchar 2006). Diabetes is a major cause of blindness, kidney failure, heart attacks, stroke and lower limb amputations. In 2012, 2.2 million deaths were attributable to DM (Mathers and Loncar 2006). In 2015, an estimated 1.6 million deaths were directly linked to DM. Almost half of all deaths attributable to high blood glucose occur before the age of 70 years. WHO predicts that DM will be the seventh leading cause of death in 2030 (Mathers and Loncar 2006).

Management of DM is intended to improve control of blood glucose levels and eliminate the possibilities of the development of chronic complications (DeFronzo 2010). Controlled diets, reduction of weight, and regular moderate exercises are lifestyle interventions designed to improve sensitivity and or responsiveness of tissue to insulin (Inzucchi 2002). The lifestyle measures should be coupled with pharmacological management as most patients fail to achieve adequate control of blood glucose with lifestyle interventions alone (Fowler 2007).

Although pharmacological treatment options are available, they are expensive. In Zimbabwe the monthly cost for type 1 diabetes management is US$400. This is unaffordable for most people in developing countries (Joseph and Jini 2011). The pathological complications caused by lack of pharmacological management often need additional drugs and are also unaffordable to many. Alternative, cheaper and more effective drugs or compounds need to be identified and tested. Plants which have a variety of phytochemical compounds with different pharmacological activities provide a good source of alternatives for testing, and it is anticipated that after scientific studies, herbal medicines and nutraceuticals could possibly form the overall treatment plan for DM, particularly in developing countries.

More than 1000 plant species are used against diabetes as folk medicine (Marles and Farnsworth 1994). Many of the plants have been evaluated for their antidiabetic activity. The antidiabetic activity of some herbal medicines is attributed to the presence of various compounds including flavonoids, terpenoids, coumarins, and phenolic compounds (Jarald et al. 2008; Tafesse et al. 2017). Investigations must be carried out to identify the chemical constituents responsible for the antidiabetic activity of the medicinal plants.

The genus *Vitex* includes shrubs and trees. The fruit of *Vitex payos* when ripe, consists of a hard nut on which the fleshy pulp rests. The thin peel layer, which is readily removable, covers the pulp. The pulp, which constitutes over 50% of the fresh mass, is composed largely of carbohydrate and fibre (Ondachi 2002). The fibre content is generally double that of carbohydrate explaining the low energy content of the pulp. Some plants in the *Vitex* genus have been studied in vitro and reported to have anti-diabetic activity (Falguni et al. 2017; Ezekwesili et al. 2012). In the work reported here the objective was to determine the effect of extracts of the fruit of *Vitex payos* on hyperglycemia in alloxan-induced diabetic rats and normal rats.

2. METHODS

2.1. Collection and Identification of Fruit

Ripe fruits of *V. payos* were collected from Nehumba Village in Rusape, Zimbabwe. The fruits were washed with distilled water and stored in a freezer at -20ºC until required. Samples of fruits were identified and authenticated by a botanist from the National Botanic Gardens of Zimbabwe.

2.2. Preparation of Samples

Frozen fruits of *V. payos* were allowed to thaw at room temperature overnight. Peels were removed by hand and the soft fleshy pulp was removed from the stone by scrapping with a knife. The pulp was dried to constant weight in the oven at 40ºC. The dried material was ground to a fine powder which was passed through a 60 mesh sieve before use. The dry powder was stored in a brown bottle at room temperature until required.

2.3. Extraction of Phytochemicals

Powdered pulp of *V. payos* fruit was hydrated with water to 0.27 g/mL. Water soluble compounds were extracted by shaking the mixture (Thermolyne Rotomix shaker M50820, Iowa, U.S.A.; 250rpm) at room temperature for 72 hours. The mixture was centrifuged at 3500rpm for 15 minutes and the supernatant was collected while the pellet underwent re-extraction and centrifugation as described. The collected supernatants were combined and freeze dried using a Christ Alpha 2-4 freeze drier (Dorfein, Germany).

2.4. Preparation of Rat Feed

A formulation for a high fat and high sugar diet for feeding the rats to pre-dispose them to DM was adapted from Taderera et al. (2014). The feed was prepared by mixing ground standard rat feed (Profeeds, Zimbabwe, 48%) as the carbohydrate, soya bean oil (8%) as the unhealthy oil and condensed milk for provision of high sugar (44%) (Taderera et al. 2014). The components were mixed and water (400 ml) was added to combine the ingredients. The mixture was moulded into balls by hand and baked at 100°C until dry. Dry feed was broken to manageable size for the rats to be able to consume.

2.5. Preparation of the Rats

Normal healthy male Sprague Dawley rats aged 4 weeks were housed under standard environmental conditions at 25°C with light and dark alternating at 12 hour periods in the animal house at the University of Zimbabwe. The initial average weight of rats was 180 ± 20g. The animals were randomly selected, marked to permit individual identification and were kept in clean cages for 5 days prior to the start of an experiment in order to acclimatize the animals to laboratory conditions. Twelve groups of five rats were used. Six groups were fed with standard rat pellets (Profeeds, Zimbabwe) whilst the other 6 groups were fed with the high fat and sugar formulated diet. Both water and food were provided *ad libitum* for all the groups.

Table 1. Grouping of the rats

Group	Treatment	
	Diabetic Rats	**Normal Rats**
I	800mg/kg	Distilled water
II	400mg/kg	800mg/kg
III	100mg/kg	400mg/kg
IV	Glibenclamide at 0.2mg/kg	100mg/kg
V	Distilled water	

Alloxan induced diabetic rats were grouped into 5 groups depending on treatment, to monitor the effect of different concentrations of aqueous extracts of *V. payos* and normal rats were grouped into 4 groups.

2.5.1. Induction of Diabetes and Grouping of Rats

DM was induced in five groups of rats with a single intraperitoneal injection of 120 mg/Kg alloxan monohydrate (Sigma –Aldrich Chemie, Steinheim, Germany) dissolved in distilled water (Madhuri et al. 2017). After one week, blood was drawn using a tail vein puncture and glucose concentrations were determined using a Direct and Quick Care® glucometer. Glucose concentrations of 11 mmol/L or above were considered as diabetic state (Taderera et al. 2014).

2.6. Administration of Extracts

Crude aqueous extracts of *V. payos,* glibenclamide or water were administered once daily using the oral gavage method to mimic the intended route of administration in humans. Glucose levels were measured using a Direct and Quick Care® glucometer, at the time of administration and every 3 days after that at the same time (9 am in the morning) for 16 days. Rat body weight was measured weekly.

2.7. Oral Glucose Tolerance Test

A glucose tolerance test was carried out as previously described by Agrawal et al. (2013) and Musabayane et al. (2005). Glucose responses were measured in normal and alloxan-induced diabetic rats. The rats were grouped into normal untreated control, normal treated and diabetic treated. An 18 h fast was allowed for all animals, before measurement of blood glucose at the start to obtain a baseline reading. Animals were loaded with glucose by oral galvage at a rate of 3 g/kg body weight. Rats were administered with either distilled water which acted as the negative control, *V. payos* extract, or glibenclamide which was the positive control. Glucose levels were measured using Direct and Quick Care® glucometer at 30 minute intervals for 2 hours.

2.8. Statistical Analysis

Statistical analysis of the data was done using Graphpad Prism 5.03 software. Unpaired t-test and one way analysis of variance (ANOVA) was done. Comparisons between means were done using Bonferroni's multiple comparison test and significant difference between the means was accepted at $p < 0.05$.

3. RESULTS

3.1. Feeding of Rats with High Fat and Sugar Diet

The mass of rats increased over the 28 days of feeding (Figure 1). The initial average weight of rats was 180 ± 20g for both groups. The average weight of the rats at 28 days was 251 ± 34g for those fed on the high fat and sugar diet and 196 ± 13g for those fed on standard rat feed. This difference was statistically significant (unpaired t-test; P value was <0.0001).

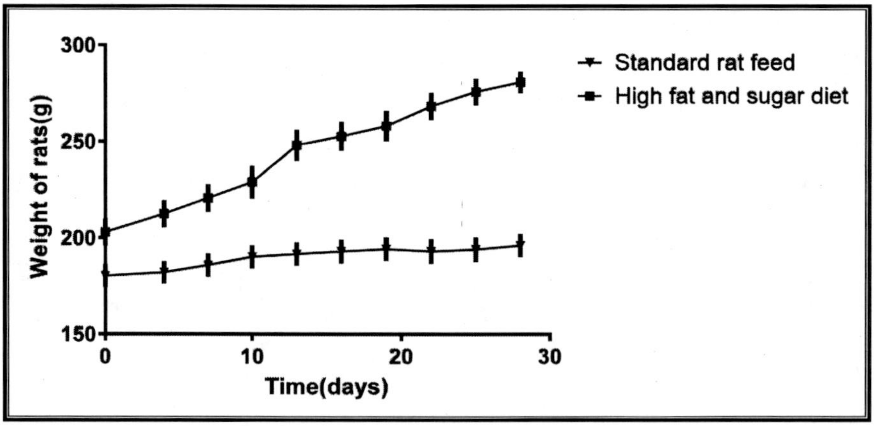

Figure 1. Mean body weights of rats that were fed with standard rat feed and rats fed with a high fat and sugar diet over 28 days. Results are the mean weight for 30 rats ± SD.

3.2. Levels of Blood Glucose in Diabetic Rats Exposed to Different Treatments

Rats in the negative control group given distilled water, died on the third day of treatment before the second measurement of blood glucose. In the group treated with glibenclamide levels of glucose decreased on the 4th day of drug administration and continued to decrease over the 16 days of treatment (Figure 2). The glibenclamide treated group showed a change in glucose levels from 12.5 ± 1.6 mmol/L on day one to 5.2 ± 0.3 mmol/L on day 16. In the groups that were treated with crude aqueous extracts of *V. payos*, a dose dependent decrease in blood glucose levels was observed with the *V. payos* treatment group of 800mg/kg having the highest glucose lowering effect from an average glucose level of 15.7 + 1.3 mmol/L reduced to 5.2 mmol/L over the 16 day period. Rats administered with 400 mg/kg extract of *V. payos* had an average glucose level of 13.8 + 1.7 mmol/L reduced to 5.9 ± 1.2 mmol/L over the 16 day period. Rats administered with 100 mg/kg extract of *V. payos* had an average glucose level of 12.1 + 0.6 mmol/L reduced to 6.9 + 1.7 mmol/L over the 16 day period.

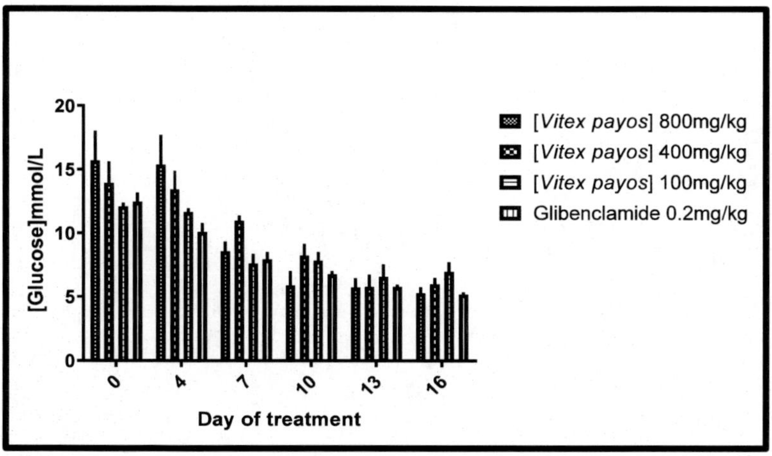

Figure 2. Levels of blood glucose in diabetic rats that were treated with crude water extracts of *V. payos* fruits and those treated with the standard drug glibenclamide. Results are given as mean ± SD for the five rats in each group.

3.3. Levels of Blood Glucose in Normal Rats Exposed to Different Treatments

As shown in Figure 3, there was no significant difference (P >0.05) in blood glucose levels of normal rats that were given water and those that were administered with different concentrations of the crude extracts of *V. payos* fruits. All treatment groups maintained constant blood glucose levels over the 16 day period.

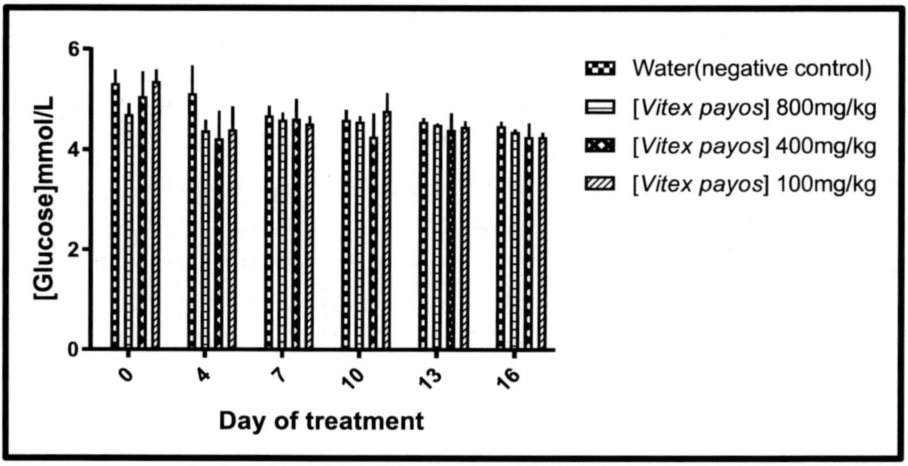

Figure 3. Blood glucose levels of normal rats that were given crude water extracts of *V. payos* and those that were given water for 16 days. Results are given as mean ± SD for the five rats in each group.

3.4. Glucose Tolerance of Diabetic Rats

An initial increase in blood glucose level was observed for all rats that were administered with extracts of *V. payos* after the initial glucose feed (Figure 4). After 60 minutes all the groups administered with extracts of *V. payos* had glucose levels gradually decreasing. Groups administered with *V. payos* at 100 mg/kg had the greatest glucose lowering effect after

120 minutes with last glucose level (5.8 + 0.3 mmol/L) close to that of glibenclamide (5.6 + 0.4 mmol/L), (P < 0.05). In rats that were treated with glibenclamide a slight increase in glucose level was observed over 30 minutes of the initial glucose feed after which glucose levels decreased.

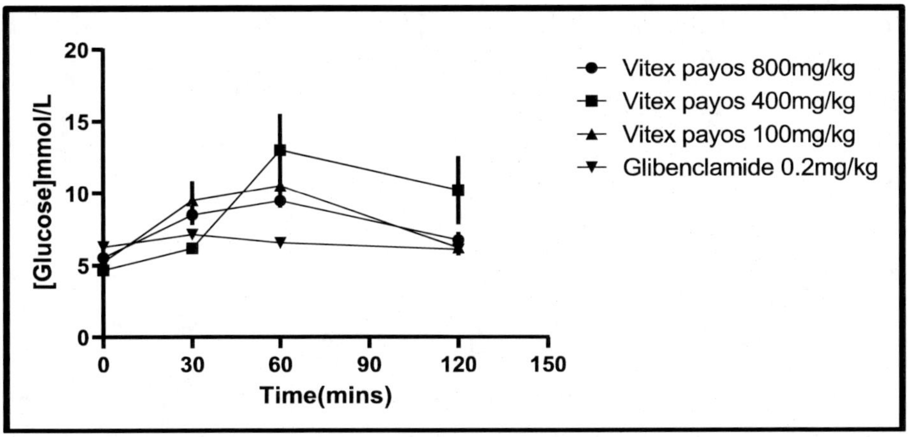

Figure 4. Comparison of glucose tolerance of variously treated diabetic rats. Results are given as mean ± SD for the five rats in each group.

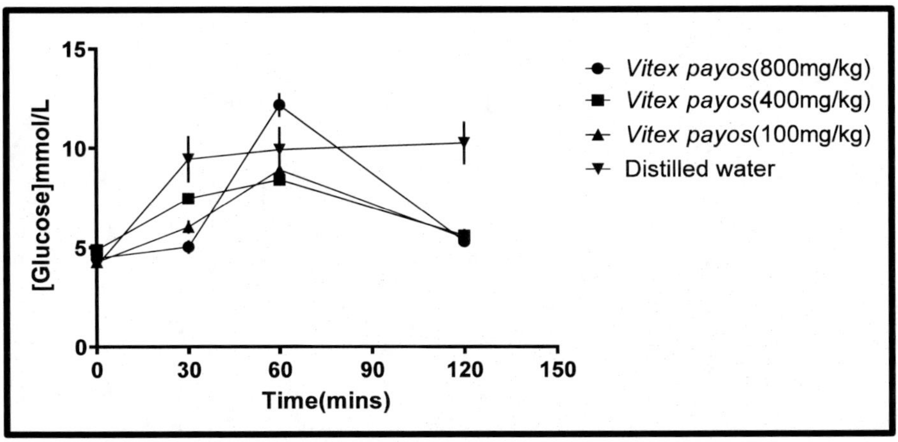

Figure 5. Comparison of glucose tolerance of variously treated normal rats. Results are given as mean ± SD for the five rats in each group.

3.5. Glucose Tolerance of Normal Rats

As shown in Figure 5, all normal rats that were administered with extracts of *V. payos* after pre-loading with 3mg/kg glucose had an initial increase in blood glucose levels over the first 60 minutes after which the glucose levels decreased. Rats that were given distilled water (negative control group) after an initial glucose feed had glucose levels continually increasing until the end of the 120 minutes.

4. Discussion

Water was used as an extraction solvent as traditional healers use water as a solvent for extraction of traditional medicine thus water was the immediate solvent of choice in the design of the experiment. Extraction with water is the intended preparation method for the juice of *V. payos* to be prepared which will be consumed by diabetic patients as anti-diabetic phytomedication.

Induction of DM, achieved using the specific diabetogenic alloxan monohydrate, resulted in 78% of the rats developing alloxan induced DM, a result similar to the 75% reported by Etuk and Muhammed in 2010. After alloxan-monohydrate induction, rats had blood glucose levels that ranged from 11-16 mmol/L. Urine of the rats had a strong fruity smell which was a further indication of DM. Alloxan monohydrate is an organ specific toxicant, that destroys the β cells of the Islets of Langerhans in the pancreas. Alloxan monohydrate induces experimental DM in animals through enhancing the generation of hydrogen peroxide, as well as inhibiting the catalase enzyme activity (Rees and Alcolado 2005). The induction responsive rate of the rats that were used was highly probable to occur as a result of the initial supplementation of the rat feed with a moderately high fat and high sugar diet to predispose the rats to obesity (Rees and Alcolado 2005). The results reported here are consistent with the fact that type II DM can be elicited by the increased fat and sugar diet that increases weight, thus interferes with insulin sensitivity (Rees and

Alcolado 2005). In the present study the consequences of high fat and high sugar diets on the body have been demonstrated. High sugar intake leads to weight gain and thus affects insulin and leptin signalling (Catoi et al. 2015). Glucose fools the metabolism in the body by turning off the body's appetite-control system. Glucose fails to stimulate insulin, which in turn fails to suppress ghrelin which is the hunger hormone, ghrelin then fails to stimulate leptin the satiety hormone which then causes one to eat more and develop insulin resistance due to obesity. Additionally, high fat and sugar consumption results in metabolic dysfunction (Catoi et al. 2015). A barrage known as classic metabolic syndrome is developed which is characterised by weight gain, obesity, decreased high density lipoprotein (HDL) and increased low density lipoprotein (LDL), elevated blood glucose levels, elevated triglycerides as well as high blood pressure (Catoi et al. 2015).

Death of diabetic rats that were administered with distilled water could be attributed to hyperglycaemia intoxication, glucose mediated neurotoxicity and metabolic acidosis (Kruyt et al. 2010). Before death, rats also exhibited giddiness, tiredness, shortness of breath and lastly death occurred which could be attributed to intoxication.

Different concentrations of crude extracts of fruits of *V. payos* were effective in lowering blood glucose levels of the alloxan-induced diabetic rats to normal levels. The extent of decrease in blood glucose levels after treatment gives an indication of whether treatment can potentially cause hypoglycaemia, a common side effect of most available drugs, which increases complications associated with DM (Bodmer et al. 2008). As shown in Figure 4, treatment of the diabetic rats with the plant extracts using different concentrations did not show excessive hypoglycaemia.

In normal rats, the maintenance of glucose at a normal 4-6 mmol/L after administration of the crude extracts of *V. payos* or distilled water (control group) can be attributed to normal blood homeostatic function of the rats as the pancreas was fully functional and insulin was being secreted normally.

Oral glucose tolerance testing indicates glucose metabolism in a short space of time without necessarily distinguishing between insulin resistance and reduced capacity of the pancreas (Etuk and Muhammed 2010). From the oral glucose tolerance tests hypoglycaemic activity in both normal and diabetic rats showed the effectiveness of aqueous extracts of *V. payos* as potential anti-diabetic agent. Administering both normal and diabetic rats with crude water extracts of fruits of *V. payos* had an effective mechanism of glucose tolerance as well as glucose lowering activity. The initial increase in blood glucose levels of normal treated rats over 60 minutes after glucose loading was not as pronounced when compared to the alloxan-induced diabetic rats. Differences in hyperglycaemic increase could probably be because the normal rats had sufficiently functional beta cells that managed to control a homeostatic balance to prevent excessive increase in blood glucose level. Both groups of normal and alloxan-induced diabetic rats that were administered with extracts from *V. payos* had their glucose levels decreasing after 60 minutes of glucose loading compared to those that were administered with distilled water (control rats) whose glucose levels continued to increase, indicating inability of the control rats to sufficiently clear the glucose, a result similar to what has been reported in other studies (Musabayane et al. 2005; Nair et al. 2006). Action of glibenclamide could be because chemical induction of diabetes altered insulin release as a result of the destroyed pancreatic islets. Glibenclamide is currently used for treatment of type II DM, as a sulfonylurease and its mechanism of action involves binding to specific receptors on the β-cells of the pancreas, resulting in shutting of potassium ATP channels. As a result, calcium channels open, leading to an increase in cytoplasmic calcium that stimulates insulin release (Taderera et al. 2014).

CONCLUSION

The crude aqueous extracts of fruits of *V. payos* have been shown to reduce blood glucose levels in alloxan- induced diabetic rats to normal levels. The aqueous extract therefore has potential for use as hypoglycaemic agent that can be used in prophylactic treatment of hyperglycaemia. Further work is required in order to isolate the potential antidiabetic compounds in fruits of *V. payos* and determine the mechanism of action of the extracts in lowering blood glucose levels.

ACKNOWLEDGMENTS

We would like to acknowledge funding from TWAS grant number17-451 RG/CHE/AS-1 and University of Zimbabwe research board.

REFERENCES

Agrawal, R. P., Sharma, N., Rathore, M. S., Gupta, V. B. Jain, S., Agarwal, V. and Goyal, S. 2013. Non-invasive Method for Glucose Level Estimation by Saliva. *Journal of Diabetes and Metabolism* 4: 1-5. doi:10.4172/2155-6156.1000266.

Bodmer, M., Meier, C., Krähenbühl, S., Jick, S. S. and Meier, C. R. 2008. Metformin, Sulfonylureas, or Other Antidiabetes Drugs and the Risk of Lactic Acidosis or Hypoglycemia: A nested case-control analysis. *Diabetes Care* 31(11): 2086–91. doi: 10.2337/dc08-1171.

Catoi, A.F., Parvu, A., Muresan, A., Busetto, L. 2015. Metabolic Mechanisms in Obesity and Type 2 Diabetes: Insights from Bariatric/ Metabolic Surgery. *Obesity Facts* 8:350-63. https://doi.org/10.1159/ 000441259.

DeFronzo, Ralph A. 2010. "Overview of Newer Agents: Where Treatment is Going." *The American Journal of Medicine* 123: S38-48. doi.org/10.1016/j.amjmed.2009.12.008.

Etuk, E. and Muhammed, B. 2010. Evidence Based Analysis of Chemical Method of Induction of Diabetes Mellitus in Experimental Animals. *Asian Journal of Experimental Biological Science* 1 (2): 331–6. https:// www.researchgate.net/publication/49599336.

Evcimen, N. D. and King. G. L. 2007. The role of protein kinase C activation and the vascular complications of diabetes. *Pharmacological Research* 55 (6): 498-510. doi: org/10.1016/j.phrs.2007.04.016.

Ezekwesili, C. N., Ogbunugafor, H. A. and J. O. Ezekwesili–Ofili, O. 2012. Anti-diabetic Activity of Aqueous Extracts of *Vitex doniana* Leaves and *Cinchona calisaya* Bark in Alloxan-Induced Diabetic Rats. *International Journal of Tropical Disease and Health* 2(4): 290-300. www.sciencedomain.org.

Falguni, M. F. Z., Islam, A., Hasan, M., Mousum, M. H. and Khatun, A. S Antioxidant and Antidiabetic Properties of *Vitex nigundo* L. Leaves. 2017. *American Journal of Life Sciences*: 5(1): 21-26. doi: 10.11648/j.ajls.20170501.14.

Fowler, Michael J. 2007. Diabetes Treatment Part 1: Diet and exercise. *Clinical diabetes* 25 (3): 105-9.

International Diabetes Federation. (015). *Diabetes: International Working Group on the Diabetic foot*. 2-4.

Inzucchi, Silvio E. 2002. Oral Antihyperglycemic Therapy for Type 2 Diabetes. *Journal of the American Medical Association* 287 (3): 360-72. doi:10.1001/jama.287.3.360.

Jarald, J. and Dharam, C. J. 2008. Diabetes and herbal medicines. *Iranian Journal of Pharmacology and Therapeutics* 7 (1): 97-106. http://ijpt.iums.ac.ir.

Joseph, B. and Jini, D. 2013. Antidiabetic Effects of *Momordica charantia* (bitter melon) and its Medicinal Potency. *Asian Pacific Journal of Tropical Diseases* 3 (2): 93–102.

Katerere, D. R. and Eloff, J. 2006. Management of Diabetes in African Traditional Medicines, In *Traditional Medicines for Modern Times: Antidiabetic Plants* edited by Soumayanath Amala, 99–116. USA: CRC Press.

Kruyt, N. D., Biessels, G. J., Devries, J. H. and Roos, Y. B. 2010. Hyperglycemia inacute ischemic stroke: pathophysiology and clinical management. *Nature Reviews Neurology* 6 (3): 145–55. doi: 10.1038/nrneurol.2009.231.

Madhuri, A. S. and Mohanvelu, R. 2017. Evaluation of Antidiabetic Activity of Aqueous Extract of *Mangifera Indica* Leaves in Alloxan Induced Diabetic Rats. *Biomedical and Pharmacology Journal* 10 (2): 1029-35 doi: http://dx.doi.org/10.13005/bpj/1200.

Marles, R. J. and Farnsworth, N. R. 1994. Plants as sources of antidiabetic agents. In *Economic and Medicinal Plant Research* published by Wagner, H. and Farnsworth, N. R. 64-106. London: Academic Press Ltd.

Mathers, C. D. and Loncar, D. 2006. Projections of global mortality and burden of disease from 2002 to 2030, *PLOS Medicine* 3 (11): e442. doi.org/10.1371/journal.pmed.

Misra, M. and Aiman, U. 2012. Alloxan: An unpredictable drug for diabetes induction? *Indian Journal of Pharmacology* 44 (4):538-39. doi: 10.4103/0253-7613.99348.

Musabayane, C. T., Mahlalela, N., Shode, F. O., and Ojewole, J. A. 2005. Effects of *Syzygium cordatum* (Hochst.) [Myrtaceae] leaf extract on plasma glucose and hepatic glycogen in streptozotocin-induced diabetic rats. *Journal of Ethnopharmacology* 97 (3): 485–90. doi: 10.1016/j.jep.2004.12.005.

Nair, S. A. Gopakumar, B. and, Subramoniam, A. 2006. Anti-diabetes and hypoglycaemic properties of *Hemionitis arifolia* (Burm.) Moore in rats. *Journal of Ethnopharmacology* 106: 192–7. https://eurekamag.com/pdf/004/004506899.pdf.

Ondachi, P. W. 2002. Nutritional studies of indigenous fruit trees in support of conservation. Capacity building in Forestry. *Annual Research Report* 19992001. 21. Nairobi: KEFRI.

Rees, D. A. and Alcolado, J. C. 2005. Animal models of diabetes mellitus. *Diabetic Medicine* 22 (1): 359–70. doi: 10.1111/j.1464-5491.2005.01499.x.

Safamansouri, H., Nikan, M., Amin, G., Sarkhail, P., Gohari, A. R., Kurepaz-Mahmoodabadi, M. and Saeidnia, S. (2014). α-Amylase inhibitory activity of some traditionally used medicinal species of *Labiatae*. *Journal of Diabetes and Metabolic Disorders* 13: 114. doi: 10.1186/s40200-014-0114-1.

Taderera, T., Gomo, E. and Chagonda L. S. 2014. The Antidiabetic Activity of an Aqueous Root Extract of *Annona stenophylla* Engl. and Diels in Non-diabetic Control and Alloxan-induced Diabetic rats. *Journal of Biologically Active Plants from Nature* 6: 315-22. doi: org/10.1080/22311866.2016.1234412.

Tafesse, T. B., Hymete, A., Mekonnen, Y. and Tadesse, M. 2017. Antidiabetic activity and phytochemical screening of extracts of the leaves of *Ajuga remota* Benth on alloxan-induced diabetic mice. *BMC Complementary and Alternative Medicine* 17: 243. doi: 10.1186/s12906-017-1757-5.

In: Phytochemicals
Editor: Iman Ryan
ISBN: 978-1-53615-478-8

Chapter 4

ROSMARINIC ACID: SOURCES, BIOACTIVITIES AND HEALTH BENEFITS

Carlos Martins-Gomes[1], Fernando M. Nunes[2], PhD, Ana Sampaio[1], PhD, Eliana B. Souto[3], PhD and Amélia M. Silva[1,*], PhD

[1]Department of Biology and Environment (DeBA-ECVA) & Centre for Research and Technology of Agro-Environmental and Biological Sciences (CITAB-UTAD), University of Trás-os-Montes and Alto Douro (UTAD), Vila Real, Portugal

[2]Department of Chemistry (DQ-ECVA) & Chemistry Research Center - Vila Real (CQ-VR), UTAD, Vila Real, Portugal

[3]Department of Pharmaceutical Technology; Faculty of Pharmacy, University of Coimbra, Pólo das Ciências da Saúde, Coimbra, Portugal

ABSTRACT

Rosmarinic acid (RA) was first isolated from *Rosmarinus officinalis* L., since then it has been identified in many plants within the *Lamiaceae*

---

* Corresponding Author's E-mail: amsilva@utad.pt.

family, such as species belonging to the genus *Rosmarinus*, *Thymus*, *Salvia* or *Origanum*, among others. Its biosynthesis begins from L-phenylalanine and L-tyrosine amino acids, resulting in an ester of caffeic acid. Rosmarinic acid, among the phenolic acids, is one of the most studied. *In vitro* and *in vivo* bioactivities of RA have been studied and are of interest to several fields of knowledge related to human health. RA and its derivatives offer a wide range of scientifically proven bioactivities, such as anti-oxidant, anti-inflammatory, anti-angiogenic, anti-proliferative, anti-microbial and others. In this review we aim to: i) discuss the literature results obtained in identification/quantification of RA and its derivatives in *Lamiaceae* family plants; ii) to explore the bioactivities of RA and derivatives, such as anti-inflammatory, antioxidant, anti-proliferative and anti-microbial, that are well described in various cell lines and microorganisms, with a diversity of assays, correlating the methods of analysis and the structure-dependent activities between RA and its derivatives. Beyond the health effects of RA, this chapter also aims to explore the most recent advances in the nanoencapsulation of RA focusing on human health applications. RA is chemically unstable and shows poor solubility in water which constrains its transport across biological barriers. Its loading in nanoformulations has been anticipated to preserve the phenolic acid bioactivities and to enhance its bioavailability.

Keywords: rosmarinic acid, *Lamiaceae* family, anti-inflammatory, antioxidant, anti-proliferative, antimicrobial, health benefits, nanotechnology, encapsulation

1. General Introduction

Rosmarinic acid (RA) was first isolated from the plant *Rosmarinus officinalis* (commonly known as rosemary), by two Italian scientists Scarpati and Oriente in 1958 (Scarpati and Oriente, 1958). RA is identified by the IUPAC designation: (2(*R*))-3-(3,4-dihydroxyphenyl)-2-[(*E*))-3-(3,4-dihydroxyphenyl)prop-2-enoyl]oxypropanoic acid; the molecular formula: $C_{18}H_{16}O_8$ and the molecular weight of 360.318 g/mol. Purified RA, at room temperature, is a powder of red-orange colour, showing a density of 1.547 g/cm^3, melting point from 171 to 175°C and a low solubility in water, being soluble in dimethyl sulfoxide (DMSO), ethanol and other organic solvents (Aydi et al., 2016). RA is commonly produced by some aromatic and medicinal plants, belonging to different families, such as: *Lamiaceae*, Boraginaceae, Apiaceae (Dresler et al., 2017, Petersen and Simmonds, 2003b). In the past years, a wide variety of bioactivities have been attributed to this polyphenol, such as antioxidant, anti-inflammatory, modulator of biochemical pathways, anti-viral, anti-tumoral, anti-allergic among others (Kim et al., 2015).

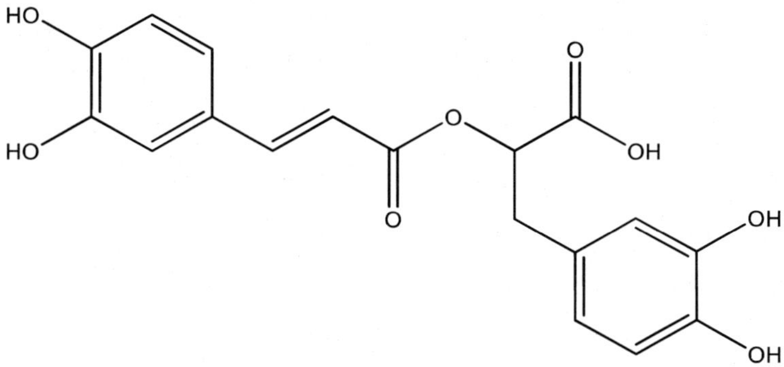

Figure 1. Chemical structure of rosmarinic acid.

2. BIOSYNTHESIS

RA (Figure 1) is an ester of caffeic acid (3,4-dihydroxycinnamic acid) and 3,4-dihydroxyphenyllactic acid (Amoah et al., 2016), being the latter commonly known as danshensu or salvianic acid A. RA was first described as a caffeic acid dimmer, but in fact the esterification is not between two molecules of caffeic acid, but rather between a molecule of caffeic acid and a molecule of salvianic acid A (Bulgakov et al., 2012).

The biosynthesis of RA and the pathways involved in its production have been studied in several plants, e.g., *Coleus blumei* (*Lamiaceae*, (Petersen et al., 1993)), *Salvia miltiorrhiza* (*Lamiaceae* (Yan et al., 2006), and others). As established for *Coleus blumei*, RA biosynthesis, from its precursor amino-acids L-tyrosine and L-phenylalanine, both aromatic amino acids, involves eight enzymes (Petersen, 2013, Petersen et al., 1993), and is summarized in Figure 2.

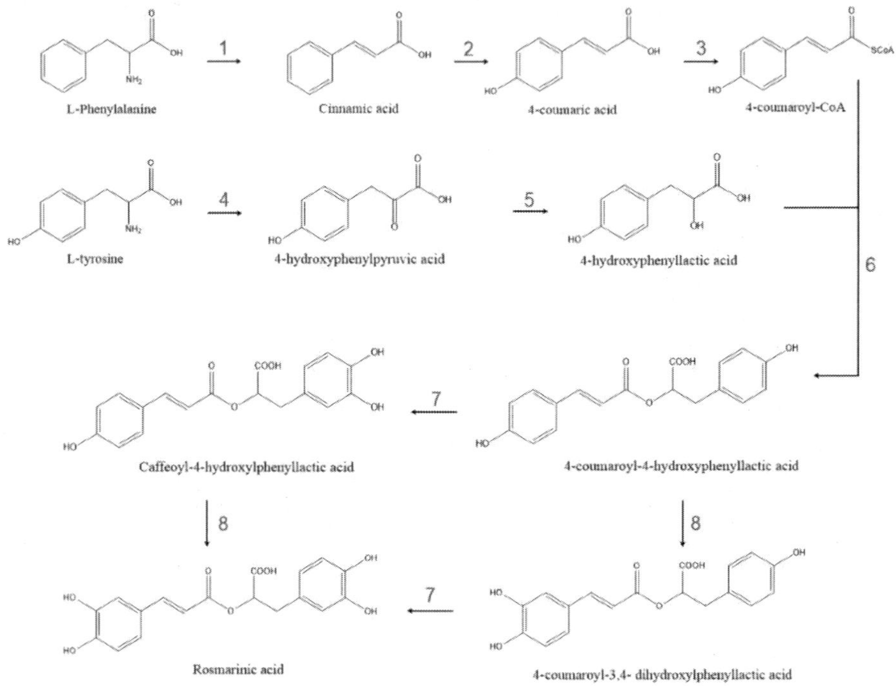

Figure 2. Rosmarinic acid biosynthesis. Biosynthesis starts with two amino acids, L-phenylalanine and L-tyrosine and involves eight enzymatic steps. Starting with the conversion of L-phenylalanine into 4-coumaroyl-CoA, by (1) Phenylalanine ammonia lyase (E.C. 4.3.1.24), (2) Cinnamic acid 4-hydroxylase (E.C. 1.14.13.11) and (3) 4-coumaric acid CoA-ligase (E.C. 6.2.1.12); and the conversion of L-tyrosine into 4-hydroxyphenyllactic acid, catalysed by (4) Tyrosine aminotransferase (E.C. 2.6.1.5) and (5) Hydroxyphenylpyruvate reductase (E.C. 1.1.1.237). (6) Rosmarinic acid synthase (E.C. 2.3.1.140) acts transferring the hydroxycinnamoyl moiety of a coenzyme A thioester to the aliphatic hydroxyl group of hydroxyphenyllactate releasing CoA. Finally, (7 and 8) hydroxyphenyllactate 3/3′-hydroxylase mediates the hydroxylation at positions 3 and 3′ of the aromatic rings (Petersen and Simmonds, 2003a, Kim et al., 2015).

Briefly, L-tyrosine is the precursor of the 3,4-dihydroxyphenyllactic acid moiety and L-phenylalanine the precursor of the caffeic acid moiety, as described in Figure 2. Both intermediary products are metabolized by rosmarinic acid synthase (6), which produces 4-coumaroyl-4-

hydroxyphenyllactic acid. RA is then obtained by the addition of two hydroxyl groups in positions 3 and 3' by hydroxyphenyllactate 3/3′-hydroxylase, addition that generates two intermediary products depending on the hydroxylation position, but both resulting in RA when the two hydroxyl groups are added (Petersen and Simmonds, 2003a, Kim et al., 2015).

RA biosynthesis occurs in a variety of plant families, from recently originated families (in an evolutionary scale) to older families originated in Cretaceous period (Kim et al., 2015), suggesting that RA plays a relevant role in plant physiology that was maintained over evolution. For example, it was suggested that RA plays a role in plant defence against herbivores and pathogens, which might be the cause of RA accumulation in some (Petersen and Simmonds, 2003a).

3. Occurrence of Rosmarinic Acid in Plants

RA has been identified and quantified in various plant species, mostly from *Lamiaceae* family (e.g., *Salvia, Thymus, Rosmarinus, Origanum*). The amount of RA in the different plants is dependent on the type of plant, but also on the extraction method. A major issue is to correlate the various quantifications, given the various factors that affect the amount of RA in each plant, such as the sampling (plant species, season and place of collection), solvent/method used which affect the extraction efficiency, quantification method, among others. Having two samples of the same species is not a guarantee that RA concentration will be identical. Climatic conditions, temperature, sun exposure, soil, season of harvest and plant portion harvested are factors that play a role in phytochemicals production by a plant, (Björkman et al., 2011). Phytochemicals represent a small percentage of a plant mass, rarely exceeding more than 10% (Harbourne et al., 2013), therefore their extraction is a key step. The first stage in any RA research would be extraction, on which the solvent/method could be limiting, as it is for any phytochemical extraction (Harbourne et al., 2013).

Table 1. RA content in mg/g dry plant (mg/g DP) in extracts obtained from plants belonging to *Lamiaceae* and Boraginaceae families

Plant	Extract type	RA (mg/g DP)	Ref.
Lamiaceae			
Salvia officinalis L.	Infusion	1.22 – 29.57	(Zimmermann et al., 2011)
Salvia miltiorrhiza Bunge	MeOH/water (30%)	2.95-7.45	(Chen et al., 2011)
Salvia officinalis L.	MeOH/water (80%)	19.50	(Bandoniene et al., 2005)
Salvia glutinosa L.		47.30	
Salvia sclarea L.		41.10	
Salvia aethiopis L.		13.30	
Salvia officinalis L.	EtOH/water (30%)	8.50-14.10	(Wang et al., 2004)
Rosmarinus officinalis L.		10.00-11.00	
Mentha spicata L.		7.1-14.3	
Lavandula angustifolia Mill.		2.0	
Melissa officinalis L.		27.4	
Prunella vulgaris L.		21.7	
Thymus vulgaris L.		4.50-8.70	
Thymus vulgaris L.	Infusion	2.85	(Martins et al., 2015)
	Decoction	2.63	
	MeOH/water (80%)	1.55	
Thymus carnosus Boiss.	Decoction	0.93	(Martins-Gomes et al., 2018)
	EtOH/water (80%)	7.18	
Thymus pulegioides L.	Decoction	8.52	(Taghouti et al., 2018)
	EtOH/water (80%)	33.47	
Thymus pulegioides L.	MeOH/water (60%)	0.89-1.4	(Boros et al., 2010)
Thymus pannonicus All.		0.95-1.27	
Thymus praecox Opiz		0.08-0.95	
Thymus glabrescens Willd		1.06-1.43	
Thymus serpyllum L.		1.12-1.37	
Thymus capitatus (L.) Hoffmanns. & Link	MeOH	1.18	(Jabri-Karoui et al., 2012)
Lavandula pedunculata (Mill.) Cav.	Infusion	11.45	(Costa et al., 2013)
	Water (*)	0.15	
	EtOH/water (*) (1:1)	6.07	

Plant	Extract type	RA (mg/g DP)	Ref.
	EtOH (*)	6.25	
Boraginaceae			
Borago officinalis L.	MeOH/water (80%)	15.00	(Bandoniene et al., 2005)
Anchusa azurea P. Mill	MeOH/water	24.00	(Dresler et al., 2017)
Anchusa undulata L.		25.90	
Pulmonaria mollis L.		36.60	
Buglossoides purpureocaerulea (L.) I.M. Johnst.		33.10	

(*) Extraction at room temperature overnight by soaking the plant material in the respective solvent.

Organic solvents (e.g., ethanol or methanol) extract a larger amount of phenolic compounds, due to the higher solubility these latter show when compared to water (Martins-Gomes et al., 2018). Thus, organic solvents provide the best extraction efficiency when combined with multiple extraction steps. Nevertheless, water-based extraction methods, such as infusions or decoctions, provide valuable information on the potential yield of RA extraction from plants when normally consumed.

Table 1 summarizes several RA levels in of different extracts (using different methods and solvents) obtained from different plant samples.

Some examples listed in Table 1 were selected to illustrate the high variability observed in RA quantifications. At first glance and only by analyzing the RA content in all the species, it is observed that the production of this phytochemical is not identical for all plants within *Lamiaceae* or Boraginaceae family.

Zimmermann et al. (2011) and co-workers studied RA levels in infusions of *S. officinalis* (commercially available as sage tea infusions) and reported high variability (RA range from 1.22 to 29.57 mg/g dry plant) which might result from the sample origin, season of collection, and of drying and conditioning methods. The same effect was observed in five samples of *S. miltiorrhiza* collected from different places in China (RA: 2.95 to 7.45 mg/g dray plant; (Chen et al., 2011)) and in various samples of *S. officinalis* (RA: 8.50 to 14.10 mg/g dry plant; (Wang et al., 2004)) both extracted with methanol (30%). Regarding the various species within the same genus, Bandoniene et al. (2005) reported the quantification of RA in various *Salvia* species extracted with a 80% methanol/water solution. Within the same genus, and without the extraction method variable, it was observed that the amount of RA in *S. glutinosa* (RA: 47.30 mg/g dry plant) was ~3.56 times higher than in *S. aethopis (*RA: 13.30 mg/g dry plant) (Bandoniene et al., 2005). The same pattern was equally observed in *Thymus* species. With an exhaustive hydroethanolic extraction, *T. carnosus* and *T. pulegioides* exhibited the higher RA concentration of the *Thymus* species listed in Table 1. *T. pulegioides* aqueous decoction even extracted more RA than *T. carnosus* hydroethanolic extract, clearly indicating the differences between the phenolic acid levels in these plants (Martins-Gomes et al., 2018, Taghouti et al., 2018) which reflect a species effect as they were collected in the same region. Nevertheless, both species exhibit a higher content when extracted with organic solvents, as confirmed by the literature, indicating that when RA content is quantified in infusions its concentration in the plant is underestimated. In other *Thymus* species, such as *T. vulgaris* collected in different geographical regions, RA content varied from 1.55 mg/g dry plant (MeOH/water (80%); (Martins et al., 2015) to 8.70 mg/g dry plant (MeOH/water (80%); (Martins et al., 2015)). In a study oriented to the phytochemical characterization of *T. vulgaris* using three different extraction methods, namely infusion, water decoction and hydroalcoholic (methanol/water (80:20, v/v), under agitation) extractions, also using different volume of solvent to dry plant weight ratio (infusion and decoction: 200 mL water; and hydroalcoholic 2×30 mL mixture); resulted in the following rank in RA extraction efficiency: infusion > decoction > hydroalcoholic (Martins et al., 2015). These different RA yields might result from the low ratio of solvent in hydroalcoholic extraction, resulting in underestimated values due to solvent saturation, and also the effect of temperature (decoction *versus* infusion), which is known to reduce the extraction recovery if in excess (Harbourne et al., 2013). The temperature effect was also observed in a study using *Lavandula pedunculata* (Costa et al., 2013), in which the authors quantified RA from two types of water extracts, infusion and soaking overnight at room temperature, resulting in ~76-fold higher amount of RA in infusion extracts

(11.5 mg/g) than in water extracts performed at room temperature (0.15 mg/g). Costa et al., (2013) also performed ethanolic and hydroethanolic (1:1) extracts, both by plant soaking at room temperature, and observed that ethanol allowed the extraction 6.25 mg/g RA (Costa et al., 2013), and the mixture EtOH/water extracted identical amount of RA (6.07 mg/g).

In the Boraginaceae family, it is possible to find species with high RA content, comparing to the *Lamiaceae* family. However, the concentrations found in the literature also vary among different species. This is the case of *B. officinalis* which shows less than half to RA concentration of *P. mollis* (Bandoniene et al., 2005, Dresler et al., 2017).

It is evident that the quantification of RA from plant extracts is highly dependent on the extraction method and the extraction temperature. Thus, some quantification may be underestimated, and the amount of RA in the plants concerned may be higher.

4. *In Vitro* and *In Vivo* Studies

4.1. Inhibition of Biologically Relevant Enzymes

The versatility of the phenolic compound's molecular structures allows for their use in a wide variety of potential pharmaceutical applications. These applications are, in part, the result of the interactions of these compounds, such as RA, with specific key molecules, such as enzymes, modulating their native activity. Given the increasing number of publications regarding RA-containing extracts and of the RA alone effect on several biological relevant enzyme inhibitions, we will discuss the most relevant.

A potential application of RA and RA-containing plants/extracts) arises from its ability to regulate the bioavailability of sugars, a relevant activity in the control/treatment of diabetes and obesity. This property is achieved by inhibiting the digestive enzymes responsible for the carbohydrate breakdown (e.g., α-amylase and α-glucosidase) and thus reducing the amount of available sugars for absorption. By inhibiting α-amylase (the enzyme that hydrolyses starch into maltose and glucose), or α-glucosidase (the enzyme that hydrolyses oligo- or disaccharides into glucose) the extracts exhibit anti-diabetic activity, reducing glucose release and absorption and thus promoting an anti-hyperglycaemic effect (e.g., (Zhu, 2015). There are several studies reporting the inhibitory effect of RA-rich extracts in both enzymes: Extracts from *Salvia elegans*, *Salvia greggii* and *Salvia officinalis* showed greater capacity to inhibit α-glucosidase than α-amylase (Pereira et al., 2018b). Extracts from *Rosmarinus officinalis* (Naimi et al., 2017) or *A. sativum*, *A. ascalonicum* (Moradabadi et al., 2013), also exhibited α-glucosidase inhibitory capacity. *T. pulegioides* decoction and hydroethanolic extracts, although rich in RA, are unable to inhibit α-amylase and can only inhibit 10% of α-glucosidase activity, presenting poor anti-diabetic activity (Taghouti et al., 2018). Contrarily to *T. pulegioides*, Alu'datt et al. (2016) reported the good efficiency of various *T. vulgaris* extracts in inhibiting both α-amylase and α-glucosidase (~40% of enzyme activity), depending on the solvent used in the extraction process. When studying an extract capacity to inhibit a specific enzyme, the effect of all its compounds must be taken in account. Although RA can inhibit these enzymes, it is possible that, due to synergistic or antagonistic effects with other phytochemicals present in the extracts may alter RA action. In a rat *in vivo* model of diabetes, RA was shown to ameliorate hyperglycaemia and insulin sensitivity (Runtuwene et al., 2016). These evidences support the use of this phenolic acid in traditional medicine, in order to improve glycaemic levels in diabetic patients, through the consumption of plants with high concentrations of RA. The RA effect in reducing glucose absorption may also be used in weight control diets.

Topal and Gulçin (2014) described RA as an efficient inhibitor of two isoenzymes of carbonic anhydrase, an essential enzyme in regulating physiological pH and CO_2 transport. The authors relate its significant inhibition to the presence of four hydroxyl groups (-OH) in RA structure, acting as functional groups and actively exhibiting non-competitive inhibition in both enzyme with low IC_{50}'s (8 μM and 47 μM).

RA is a potentially interesting drug molecule for conditions involving skin pigmentation, such as in the treatment of vitiligo, where a hypopigmentation is observed in some parts of the skin. RA was shown to induce melanogenesis in a protein kinase A dependent mechanism, which increased the expression of tyrosinase, a key enzyme in melanin production, with dose-response, to concentrations up to 100 μM (Lee et al., 2007). Fujimoto et al. (2010), on the other hand, describe RA as a tyrosinase inhibitor, with an IC_{50} of 1.40 mM. Contrarily to Lee et al. (2007) who carried out the assays using cell culture exposed to RA, observing tyrosinase expression as the result of RA induction of specific metabolic pathways (CREB (cAMP response element-binding protein) and PKA (protein kinase A)), leading to increased melanin content resulting in skin pigmentation, Fujimoto et al. (2010) used standard *in vitro* enzymatic inhibition assay, with isolated mushroom tyrosinase. These results are however contradicted by Lee et al. (2007). This would result in a loss of melanin in the skin, resulting in hypopigmentation. Although both studies are very different regarding the methodologies, further studies are needed to ascertain the RA action as a skin pigmentation modulator.

The action RA-rich extracts, in the modulation of enzymes responsible for the neurotransmitter degradation after exocytosis, such as acetylcholinesterase, (AChE), have been studied as a potential source of therapeutic drugs to prevent/treat several neurodegenerative diseases. Most neurodegenerative diseases are a consequence of a decline in neurotransmitter production, as is the case of Alzheimer's disease, Parkinson's disease, and senile dementia, in which decreased acetylcholine levels in the brain are the major common biochemical change. Restoring acetylcholine levels is pharmacologically achieved using drugs that inhibit the AChE and the butrylcholinesterase (BChE), enzymes that are involved in the termination of neurotransmission process, by hydrolysing acetylcholine. Although, having the same activity, they are found in different tissues, AChE mainly in chemical synapses and red blood cells and BChE in the plasma. RA-rich extracts have been reported to inhibit these enzymes.

Gülçin et al. (2016) reported the high capacity of RA in inhibiting acetylcholinesterase (with K_i value of 42.52 pM) and butyrylcholinesterase (with K_i value of 121.60 pM), and suggested these compounds as having ability to revert the symptoms of pathologies such as Alzheimer's disease *T. pulegioids* aqueous and hydroetanolic extracts also showed great capacity to inhibit AChE (~80% inhibition of enzyme activity at 0.5 mg/mL of extracts), being referred as having neuroprotection potential (Taghouti et al., 2018).

4.2. Antioxidant Activity

One of the most commonly described uses of aromatic and medicinal plants is the ability to reduce oxidative stress. This antioxidant activity has been intensely described in *Lamiaceae* family plants. With even more relevance, is the association of the antioxidant potential with specific compounds, as it is the case of RA. In various *Lamiaceae* plants, as marjoram, rosemary, sage or thyme, RA was described as a contributor for the high antioxidant activity attributed to these species (Embuscado, 2015).

Regarding phenolic compounds, the number and distribution of hydroxyl groups, directly affects its antioxidant potential. As a phenolic acid, RA scavenging is likely to be exhibited by trapping free radicals. Its structure (Figure 1), with four phenolic hydroxyl groups, two in each aromatic ring, allows RA to trap metals (by chelation) and radical trough the donation of hydrogen atoms ($H^•$) (Brewer, 2011).

RA ability to reduce oxidative stress is well described, and can be subdivided in two major mechanisms, depending on the methods of testing: standard *in vitro* testing and direct application in biological models. In the first topic, standard assays are used to assess a specific extract or isolated compound scavenging activity. This is the case of $DPPH^•$, superoxide radical and β-carotene bleaching, used to demonstrate that this phenolic acid has high efficiency in scavenging free radicals (Terpinc and Abramovič, 2010, Alamed et al., 2009). Like most phenolic compounds, RA does not produce 100% antioxidant activity at all concentrations. Muñoz-Muñoz et al. (2013) verified that in concentrations up to 13.5 mM, H_2O_2 is generated by RA pro-oxidant activity, and above this concentration the antioxidant effect starts to act, decreasing H_2O_2 concentration. Thus, at higher RA concentrations, hydrogen peroxide is eliminated, but at lower concentrations RA induces H_2O_2 production.

However, not only by scavenging free radical, RA can act as an oxidative stress reducer modulating the expression of enzymes involved in oxidative stress defense. RA was described to induce oxidative stress enzymes expression in Human keratinocytes cell line (HaCaT,), such as superoxide dismutase (SOD), catalase (CAT) or heme-oxygenase (HO-1), which are deeply related to oxidative stress balance, through the activation of Nuclear factor-like 2 (Nrf2) (Fernando et al., 2016). These enzymes control the uprising of free radicals' concentration in the cells, and thus preventing the undesirable oxidation of DNA, lipids or proteins. At intracellular level, RA efficiently reduced free radical concentration in cells exposed to hydrogen peroxide as well as it improved cell resistance to UVB-induced oxidative stress (Fernando et al., 2016). The expression of these proteins, as well as glutathione peroxidase (GPx), were also reported in an *in vivo* model (mice model) exposed to RA (Zhang et al., 2015).

Another relevant activity, regarding oxidative stress, is the capacity to prevent lipid peroxidation. Fadel et al. (2011), described RA as a molecule capable of being inserted in the lipid membrane, in concentrations up to 1% (% RA/lipid), without disrupting its integrity or affecting its structure. The insertion is spontaneous, with the phenolic acid being inserted close to unsaturated lipids. This fraction of RA is capable of reducing lipid peroxidation. Apart from lipids and proteins, nucleic acids are one of the major targets of ROS, resulting in devastating effects such as carcinogenesis and mutagenesis. As part of its role as scavenger, RA has proven to protect DNA from harmful oxidation by hydrogen peroxide (Sevgi et al., 2015). Thus, RA acts as a scavenger against ROS, protecting and preventing oxidation of the major targets of these reactive species, i.e., proteins, lipids and DNA. This has been demonstrated not only by standard chemical methods, but also by experiments carried out in biological systems and described in this section (4.2). Given the big number of plants that produce and accumulate high concentrations of RA, it is understandable the high number of publications reporting the added-value of medicinal plants in preventing oxidative stress events. However, oxidative stress cannot be considered as an isolated event, since it has deep connections with the inflammatory processes. The reactive nitrogen species (RNS) nitric oxide (NO), a key molecule in the inflammatory process and also implicated in the modulation of several cell events, is efficiently scavenged by RA (Kelsey et al., 2010, Qiao et al., 2005), in a dose-dependent manner (Martins-Gomes et al., 2018).

4.3. Anti-Inflammatory Activity

The capacity of a plant or extract to prevent or treat inflammation is well associated with the *Lamiaceae* family, in part due to the high content in RA (Sotnikova et al., 2013). During the inflammatory process, various metabolic events are processed in order to generate the common symptoms: swelling, pain, increased body temperature and skin redness (Santangelo et al., 2007). Leukocytes are the major intervenient, migrating from the blood vessels to the inflamed tissue in response to specific molecules.

The anti-inflammatory activity of RA is due to its action inhibiting enzymes and specific pathways involved in the inflammatory process, such as cyclooxygenases (COX) and lipoxygenases, the complement cascade or the production of cytokines (Gamaro et al., 2011). This cascade starts with arachidonic acid which can be recruited from cell membrane or obtained from the diet, and then it is metabolized by lipoxygenases (-5, -12, -15) or COX (-1 and -2). Lipoxygenases produce hydroperoxyeicosatetraenoic acid, hydroxyeicosatetraenoic acid and leukotrienes. The products of COX-1 and COX-2 activity are prostaglandins, thromboxanes and prostacyclins. Several polyphenols, including RA have been reported as inhibitors of lipoxygenases (Chen, 2011). Some molecular targets of RA, and other polyphenols, in inflammatory cascade are shown in Figure 3.

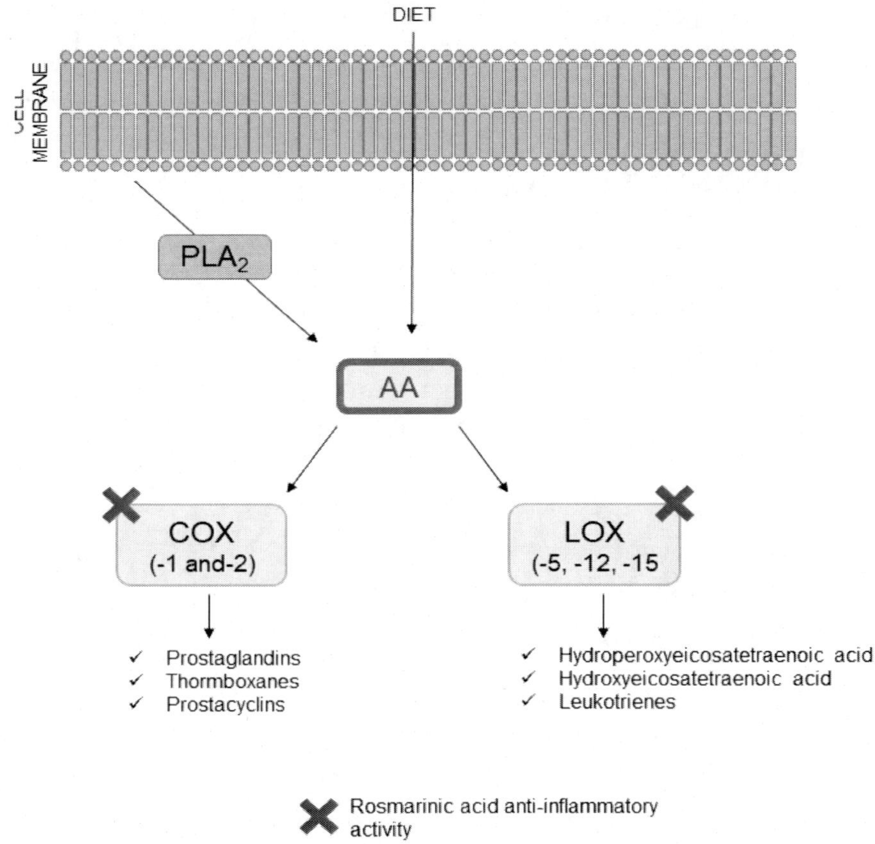

Figure 3. Arachidonic acid-dependent inflammation pathway. Potential targets of RA, as described in several reports. Adapted from (Santangelo et al., 2007, Chen, 2011, Astudillo et al., 2012).

The cytokines are pro-inflammatory polypeptides, produced when MAPK (Mitogen-activated protein kinases), JAK-STAT (Janus kinases-Signal Transducer and Activator of Transcription proteins) and NF-kB (nuclear factor kappa B) pathways are triggered. Among these are interleukin (IL) -1, -6, -18 and TNF (tumor necrosis factor), crucial molecules in the inflammatory cascade (Navarro-González et al., 2011) and thus, important targets for RA and other phenolic compounds to prevent this metabolic response.

When describing the metabolic pathways of inflammation, it is impossible not to mention the MAPK's. This group includes three major protein families: ERK (extracellular signal–regulated kinases), p38 and JNK (c-Jun N-terminal kinases). These proteins regulate the inflammatory cascade, controlling the synthesis and release of pro-inflammatory signals in the inflamed tissue (Chu et al., 2012). Chu et al. (2012) have showed that RA prevented ERK activation and, consequently, the remaining pathway.

TNF-α activation of NF-kB is a result of degradation or phosphorylation of the inhibitor of kappa B (IkB), and thus RA exerts anti-inflammatory activity through NF-kB cascade, either by preventing its activation (keeping IkB in its functional form) or by preventing p50 and p65 translocation to the nucleus (Moon et al., 2010).

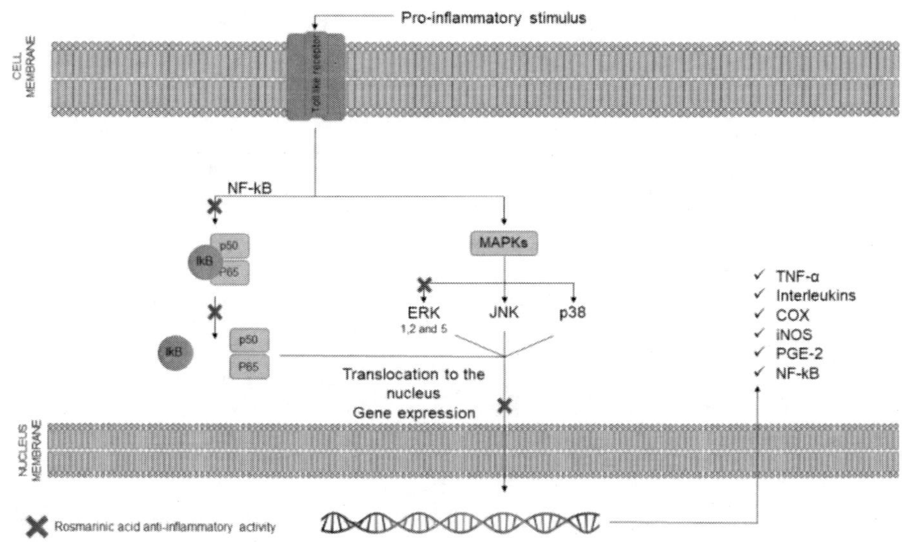

Figure 4. NF-kB- and MAPKs-dependent inflammation pathway. Potential molecular targets of RA, illustrating its anti-inflammatory activity. Adapted from (Santangelo et al., 2007, Calder, 2011, Costa et al., 2012, Arulselvan et al., 2016).

Matrix metalloproteinases (MMP) family also contribute to inflammation. These zinc-dependent endopeptidases act by increasing chemokines and cytokines in the extracellular matrix, namely IL-1β (Rahbardar et al., 2017).

As previously mentioned, NO is an active intervenient in inflammation. Apart from being a free radical, this gas is produced in high concentrations by inducible nitric oxide synthase (iNOS), through the conversion of L-arginine to L-citruline (El-Hattab et al., 2012). This makes NO an ideal marker for tracking the inflammatory process, as well as for screening the capacity of any substance to prevent inflammation. NO quantification is the most common assay to access anti-inflammatory activity, applicable to various experimental models, from *in vitro* to *in vivo*. In Table 2, the most recent studies regarding the anti-inflammatory activity of RA, are summarized. The first observation that stands out is the efficient anti-inflammatory activity of RA regardless the experimental model. In these studies, various mice and rat strains as well as several cell lines have been used. Several models have been optimized to study inflammation, with diverse inducers, allowing a deeper and comprehensive understanding of the underlying mechanisms. Due to the simple quantification of the NO production, various authors reported a decrease in NO production after RA exposure (Chen, 2011, Martins-Gomes et al., 2018), to lipopolysaccharide (LPS)-stimulated RAW 264.7 cells (macrophages). Considering all the studies shown in Table 2, several group of outcomes can be highlighted. First, NF-kB inhibition (either by non-translocation of p50 and p65 to the nucleus or by inhibition of IkBα degradation), with consequent inhibition of MAPK pathway, which decreases production of interleukins (mainly IL-1β and IL-6), TNF- α, PGE2. The second interesting outcome is the reduction of oedema and cell migration to the inflamed tissue in mice and rats (Gamaro et al., 2011, Boonyarikpunchai et al., 2014). Furthermore, the increased production of an anti-inflammatory cytokine, the IL-10, has also been reported (Mueller et al., 2010).

Reporting that RA exerts anti-inflammatory activity is more than presenting an alternative treatment for an isolated pathology. Inflammation plays a role in the outcome of various diseases, as acute lung injury (Chu et al., 2012), neuropathic pain (Rahbardar et al., 2017), rheumatoid arthritis and osteoarthritis (Mueller et al., 2010) and ischemic stroke, as an isolated event or as a result of other pathologies likes diabetes (Luan et al., 2013).

Currently, several drugs are used to treat inflammation (such as, steroidal/non-steroidal anti-inflammatories and corticosteroids), but they carry with serious adverse side-effects (Mueller et al., 2010), therefore, natural compounds such as RA may be regarded as strong alternatives against classical drugs.

Table 2. Anti-inflammatory mechanism triggered by the action of RA in several experimental models of inflammation

Experimental model	**Anti-inflammatory mechanism**	**Reference**
Inflammation inducer		
Wistar rat	Lower NO, COX-2, IL-1β, PGE-2 and MMP-2 concentration after 7- and 14-days exposure	(Rahbardar et al., 2017)
Neuropathic pain		
Human neuroblastoma SH-SY5Y cells	Inhibition of NF-kB Inhibition of p-IkB activation Lower HMGB1 expression	(Luan et al., 2013)
TNF-α		
Sprague-Dawley rat	Lower MPO activity Lower HMGB1 expression Inhibition of NF-kB	
Streptozotocin		
Wistar rat	Reduction of cell migration to the inflamed tissue	(Gamaro et al., 2011)
Carrageenan		
ICR mouse	Reduction of paw oedema	(Boonyarikpunchai et al., 2014)
PGE2		
BALB/c mouse	Reduced TNF-α, IL-1 β and IL-6 production Inhibited p-ERK activation Reduction of cell migration to the inflamed tissue	(Chu et al., 2012)
LPS		
Human monocytic leukemia cells (U937)	Inhibition of NF-kB activation Reduced expression of p50 and p65 in the nucleus Reduction of ROS Prevented IkBα degradation	(Moon et al., 2010)
TNF-α		
Human leukemic mast cells (HMC-1)	COX-2 and NF-kB inhibition	(Oh et al., 2011a)
Phorbol 12-myristate 13-acetate (PMA)		
BALB/c mouse	Reduced IgE, TNF-α, IL-1β and IL-6 production	
Ovalbumin		
Wistar rat	Reduced pro-inflammatory cytokines release Inhibition of neutrophil activation	(Rocha et al., 2015)
Carrageenin		

Experimental model / Inflammation inducer	Anti-inflammatory mechanism	Reference
Wistar rat / Streptozocin	Reduced TNF-α and IL-1β production	(Sotnikova et al., 2013)
Primary human umbilical vein endothelial cells (HUVEC) / LPS	Inhibition of (HMGB-1) release	(Yang et al., 2013)
Murine macrophages (Raw 264.7) / LPS	Lower NO concentration iNOS and COX-2 inhibition	(Shen et al., 2010)
Murine macrophages (Raw 264.7) / LPS	Increased IL-10 expression Reduced IL-6	(Mueller et al., 2010)
Murine macrophages (Raw 264.7) / LPS	Lower NO concentration	(Martins-Gomes et al., 2018)

Abbreviations: HMGB1, High mobility group box-1; MPO, Myeloperoxidase; PGE2, Prostaglandin E2; IgE, Immunoglobulin E.

4.4. Anti-Tumoural Activity

For any natural compound, the capacity to prevent carcinogenicity, mutagenicity or the growth of tumours is a desirable attribute. In this field of study, RA has a relevant role, with a good share of publications that highlight these activities.

In one of the comprehensive and detailed studies about RA interaction with cell death, Wu et al. (2015) described RA as an apoptosis inducer in tumoral cells through modulation of ERK pathway and the expression of anti-tumoral genes. The authors reported that drug resistant leukaemia cells exposed to RA exhibit morphological disturbances consistent with cell death, like swelling and the presence of apoptotic bodies. With the aim to study the cell death induction, the authors correlated the anti-tumoral activity with NF-kB deregulation, implicated in cell death, and revealed the capacity of RA to perform cell cycle arrest in Sub-G1 and G2/M (Wu et al., 2015). Using Annexin/PI double staining, RA was shown to induce both necrosis and apoptosis (Wu et al., 2015), the cell death involved DNA damage but was independent of caspases 3, 7 and 9; although the authors detected depolarization of the mitochondria membrane potential and loss of cell adhesion (Wu et al., 2015). Xu et al. (2010) have correlated RA anti-tumour activity to its efficient capacity to inhibit matrix metalloproteins expression that prevented cell migration, invasion and adhesion.

In leukaemia cells (U937, Human monocytic leukemia), Moon et al. (2010) demonstrated that RA increase TNF-α capacity to induce apoptosis, resulting in increased cell death by synergism, as comparing to TNF-α alone effect, and as RA exhibited no cytotoxicity at the tested concentrations (up to 60 μM). U937 cells treated with both agents (RA and TNF-α) showed cell cycle arrest in sub-G1 and DNA fragmentation in higher levels than the isolated effect of each one (Moon et al., 2010). Additionally, the authors correlate the observed cell death with caspase-3 activity and, once again, exclude NF-kB pathway, due to the strict control exhibited by RA.

In a mouse model for hepatocarcinoma, RA significantly reduced tumor growth, without loss of weight (typical of some anti-tumoral drugs) and its use in the treatment did not affect other organs (e.g., thymus, spleen and liver). Tumoral growth markers as inflammation and angiogenesis were also diminished (Cao et al., 2016), conferring RA the capacity to prevent tumoral growth, without the disadvantages of conventional drugs. In an assay to study the prevention of tumoral growth in mouse skin, RA effectively regulated caspase 3 and 9 values and inhibited p53 and blc-2 activation, resulting in the elimination of tumoral cells (Sharmila and Manoharan, 2012). RA is non-mutagenic and acts preventing mutagenicity (Furtado et al., 2008).

Han et al. (2015) studied the influence of RA in the Warburg effect in MKN4, a Human gastric cancer cell line. The Warburg effect refers to the ability of tumoral cells to increase the production of energy trough glycolysis by upregulating HIF-1 (hypoxia-inducible factor). This is a standard phenomenon observed in cancer development. RA exhibited cytotoxicity in MKN4 cells and decreased its glucose consumption, as well as lactate formation. *In vivo*, using BALB/c mouse model, the same effect was observed. Thus, the authors confirmed the anti-Warburg effect activity of RA, as it also inhibits HIF-1and inflammation (Han et al., 2015).

4.5. RA and Derivatives: Other Relevant Bioactivities and Applications

Apart from the above described bioactivities, RA presents a wide spectrum of applications that stretches to other areas of interest. This phenolic acid also shows antiviral (Swarup et al., 2007), antibacterial (Bais et al., 2002) and neuroprotective effects (Fadel et al., 2011). For example, RA is known to prevent β-amyloid plaques formation, by preventing β-amyloid aggregation (Fadel et al., 2011), an event associated with Alzheimer's diseases. The prevention of other pathologies, namely liver diseases (fibrosis, chronic injury or cirrhosis), have been successfully demonstrated in several experimental models. Using HSC cells (hepatic stellate cells), Li et al. (2010) evidenced the cytotoxicity induce by RA and the reduced production of α-SMA (smooth muscle actin), CTGF (connective transforming growth factor) and TGF-β1 (transforming growth factor- β 1), markers of fibrosis, as well as improved the morphology and markers of liver fibrosis in a rat model, results also supported by other authors (Domitrović et al., 2013). Within protective effects, RA was also shown to protect the kidney of mice exposed to cisplatin, a powerful drug used in the treatment of various cancer types (Domitrović et al., 2014). RA decreased the non-specific toxicity of cisplatin, by restoring oxidative stress to control levels (by controlling HO-1 (Heme oxygenase) and CYP2E1 (Cytochrome P450 2E1) expression), the inflammation cascade (via NF-kB inhibition) and the prevention of apoptosis (Domitrović et al., 2014). About the protective effect of RA against external stimulus, the ability to prevent oxidation and consequent damage in keratinocytes exposed to UVB radiation has also been described (Vostálová et al., 2010).

RA acted as a radical scavenger, reducing DNA damage and avoiding the activation of caspases (and thus apoptosis) and interleukins (Vostálová et al., 2010).

Regarding its antimicrobial activity, RA has been described as capable of preventing the growth of several fungi from the genera *Aspergillus*, *Rhizoctonia*, *Fusarium*, *Versicillium* and *Alternaria,* and fungi-like organisms such as the Oomycota *Phytophthora* and *Pythium* as well as of inhibiting the proliferation of three bacterial species: *Xanthomonas campestris*, *Pseudomonas fluoroescens* and *Erwinia carotovora* (Bais et al., 2002). The fungi species presented morphological changes within the hyphae, spores and the cell wall. RA also exhibited bactericidal effect against *Pantoea agglomerans* and *Stenotrophomonas maltophilia, Klebsiella* sp. and *Streptomyces* sp. (Wang et al., 2012). Abedini et al. (2013) reported that RA affected both gram positive and gram-negative bacteria.

RA exhibited anti-viral activity, by reducing the associated inflammation and inhibiting its proliferation, RA was appointed as a potent agent against the Japanese Encephalitis virus (Swarup et al., 2007) and also a HIV (Human immunodeficiency virus) inhibitor by inhibiting its integrase (Dubois et al., 2008).

Apart from RA, other compounds have been classified as dimmers, trimers or even tetramers of caffeic acid. Some of them are identified as rosmarinic acid derivatives, such as salvianolic acid A, salvianolic acid B and rabdosiin. Unlike RA, not much relevance has been given to its derivatives, and thus, the spectrum of know bioactivities is yet small. Table 3 summarizes the bioactivities associated with these RA derivatives.

In a generic analysis, it is possible to observe that both RA and its derivatives share common pharmacological effects, most likely due to the fact that the bioactivity could be induced by a specific functional group within the molecule. This is the case of antioxidant activity, anti-inflammatory, anti-viral and anti-tumoral, described in Table 3.

Table 3. Main bioactivities and structure of rosmarinic acid derivatives

Compound	Biological activity	Ref.
Salvianolic acid A	Anti-inflammatory activity through IKK (NF-kB pathway), iNOS and COX-2 inhibition	(Oh et al., 2011b)
	Anti-thrombotic and antiplatelet	(Fan et al., 2010)
	Apoptosis inducer in acute myeloid leukaemia	(Pei et al., 2018)
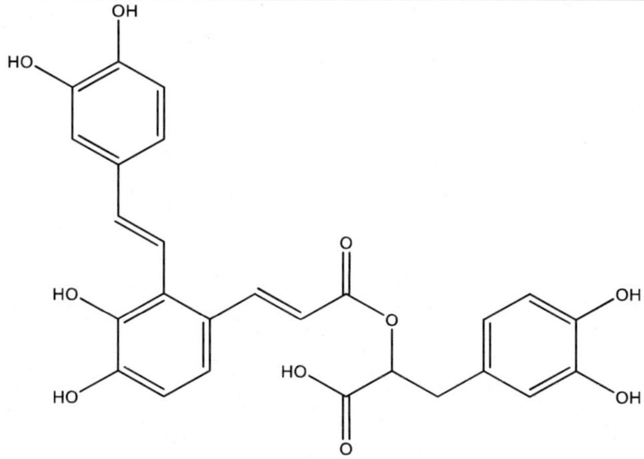		
Salvianolic acid B	Neuroprotective and anti-inflammatory; Inhibition of NO production, TNF-α, IL-1 and ROS	(Wang et al., 2010)
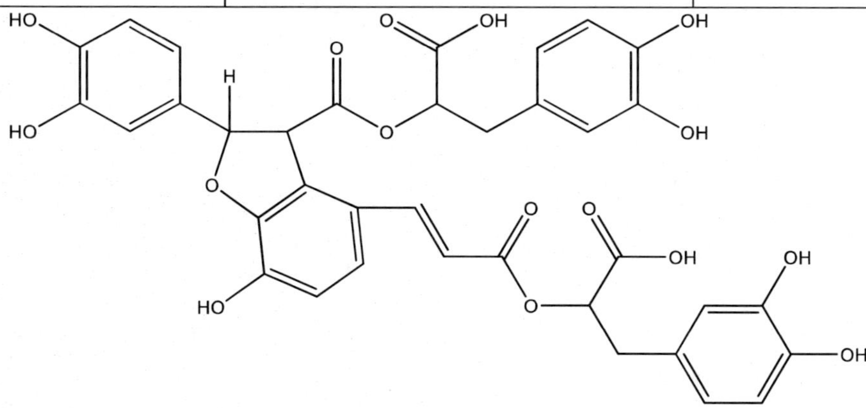		

Table 3. (Continued)

Compound	Biological activity	Ref.
Rabdosiin	Antioxidant activity	(Yoo et al., 2014)
	Anti-HIV activity Anti-glomerulonephritic activity	(Bulgakov et al., 2012)

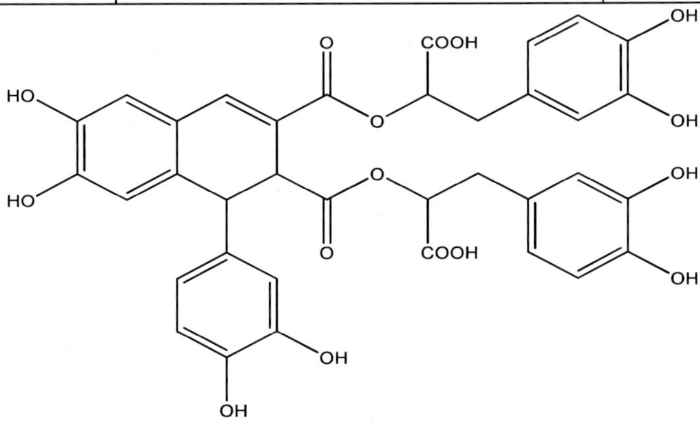

Abbreviations: IKK Inhibitor of nuclear factor kappa-B kinase (subunit beta)

5. Nanoformulations and Pharmaceutical Applications

With the aim to enhance the pharmacological properties of RA, several strategies have been employed for site-specific delivery thereby improving its bioavailability in vivo, among which nanoformulations have been receiving particular attention.

Polymeric nanoparticles composed of polyacrylamide, cardiolipin and poly(lactide-co-glycolide) (PLGA) have been surface-tailored with 83-14 monoclonal antibody and loaded with RA and curcumin (Kuo and Tsai, 2018). The authors reported the enhanced drug viability in beta-amyloid enriched SK-N-MC cells and capacity to cross the blood brain barrier. The developed system was shown to reduce the fibrillary beta-amyloid-induced neurotoxicity.

Similar results have also been reported by Kuo and Rajesh (2017). Polyacrylamide-chitosan-PLGA nanoparticles were surface-grafted with apolipoprotein E (ApoE) and cross-reacting material 197 (CRM197) and loaded with RE to be used in neurodegenerative diseases (e.g., Alzheimer's disease) (Kuo and Rajesh, 2017). The particles were tested *in vitro* (in human brain-microvascular endothelial cells, RWA264.7 cells, and in beta-amyloid-enriched SK-N-MC cells), demonstrating that trans-endothelial electrical resistance could be reduced while RA uptake increased by increasing the concentration of CRM197 and ApoE.

Chitosan is commonly used in the production of nanoparticles attributed to a range of properties including its antimicrobial activity (Khattak et al., 2019, Shariatinia, 2019). The polysaccharide has been used to prepare nanoparticles by ionic gelation using chitosan and sodium tripolyphosphate to be loaded with RA for ocular application (da Silva et al., 2016). RA-loaded chitosan nanoparticles were shown to be non-cytotoxic against retina pigment epithelium (ARPE-19) and human cornea cell line (HCE-T), with a slight enhanced apparent permeability in HCE monolayer cell line when using nanoparticles (in comparison to the free drug solution), suggesting increased retention time via mucin interaction upon application.

Salvia officinalis (sage) and *Satureja montana* (savory) extracts containing RA were loaded in chitosan nanoparticles with the aim to protect the antioxidant properties of the phenolic acid (da Silva et al., 2015). The authors reported above 50% of association efficiency for RA with higher antioxidant activity values than the extracts alone. To improve RA stability and solubility, and enhance the capacity to cross biological membranes, chitosan microparticles have also been proposed (Casanova et al., 2016).

In the production of gold nanoparticles, rosmarinic acid was used as a reducing agent to reduce gold ions into nanoparticles. Lim and Park (2018) described a green synthesis approach, highlighting the potential involvement of hydroxyl and carbonyl functional groups of rosmarinic acid in the synthesis of gold nanoparticles (Lim and Park, 2018).

Other recognized green-based nanoparticles are those composed of physiological lipids. Solid lipid nanoparticles (SLN), engineered from biodegradable and biotolerable lipids (e.g., triglycerides, fatty acids, waxes) have been reported to enhance bioavailability of natural compounds (including polyphenols) and to modify their release profile (Carbone et al., 2018, Pereira et al., 2018a, Zielinska et al., 2018).

SLN composed of Witepsol and Carnauba waxes have been loaded with RA and tested *in vitro* and *in vivo* (Madureira et al., 2016b). The cyto/genotoxic profile of RA-loaded SLN was concentration dependent, exhibiting *in vivo* safety profile in orally treated Wistar rats for 14 days (RA-loaded SLN at 1 and 10 mg/kg body weight/day).

When administered by oral route, SLN were describe to prevent the loss of RA during digestion, attributed to a slower release under simulated gastric conditions and then increasing the release in the intestine, assuring and higher rate of RA absorbed by intestinal epithelium (Madureira et al., 2016a). The efficient delivery of RA by the use of SLN could even benefit the treatment of neurological diseases. Huntington's disease is a neuro-degenerative pathology characterized by loss of cognition and neurological disturbances, with a progressive and deadly development. RA-loaded SLN have been tested by intranasal administration as a pathway to increase the brain uptake of the phenolic acid. This is based on RA ability to protect neurological cells from oxidation and undesirable apoptosis (Bhatt et al., 2015). The authors concluded that nasal delivery route may be a valid option, overcoming the need of intra-venous administration, as it delivers more RA to the brain thereby limiting the systemic distribution (Bhatt et al., 2015). The use of nanoformulations to enhance the pharmacological and therapeutic properties RA is a growing field, nevertheless still requiring further validation.

Conclusion

In conclusion, RA is present in a wide variety of plant species in high concentrations, with a well-defined biosynthesis pathway and extraction protocols. The wide range of bioactivities attributed to this phenolic acid and its low/limited cytotoxicity to healthy tissues, together with the high concentrations RA is found in commonly consumed plants, open interesting perspectives for the use of this phenolic acid as a valuable nutraceutical. RA can act in the prevention and treatment of diverse diseases, regulate the expression of several key proteins, and prevent inflammation and oxidative stress. As described in this chapter, RA can contribute to endless purposes within pharmaceutic industry, but still lacks direct application in more complex experimental models. Before its direct use as a drug, several more studies are required, but there is for sure health benefits in its consumption either as aromatic plant (as seasoning or condiment) or for medicinal purposes in the form of infusions or other preparations.

References

Abedini, A., Roumy, V., Mahieux, S., Biabiany, M., Standaert-Vitse, A., Rivière, C., Sahpaz, S., Bailleul, F., Neut, C. & Hennebelle, T. 2013. Rosmarinic acid and its methyl ester as antimicrobial components of the hydromethanolic extract of *Hyptis atrorubens* Poit. (*Lamiaceae*). *Evidence-Based Complementary and Alternative Medicine,* 2013.

Alamed, J., Chaiyasit, W., Mcclements, D. J. & Decker, E. A. 2009. Relationships between free radical scavenging and antioxidant activity in foods. *Journal of Agricultural and Food Chemistry,* 57, 2969-2976.

Alu'datt, M. H., Rababah, T., Johargy, A., Gammoh, S., Ereifej, K., Alhamad, M. N., Brewer, M. S., Saati, A. A., Kubow, S. & Rawshdeh, M. 2016. Extraction, optimisation and characterisation of phenolics from *Thymus vulgaris* L.: phenolic content and profiles in relation to antioxidant, antidiabetic and antihypertensive properties. *International Journal of Food Science & Technology,* 51, 720-730.

Amoah, S. K., Sandjo, L. P., Kratz, J. M. & Biavatti, M. W. 2016. Rosmarinic acid–pharmaceutical and clinical aspects. *Planta Medica,* 82, 388-406.

Arulselvan, P., Fard, M. T., Tan, W. S., Gothai, S., Fakurazi, S., Norhaizan, M. E. & Kumar, S. S. 2016. Role of antioxidants and natural products in inflammation. *Oxidative Medicine and Cellular Longevity,* 2016.

Astudillo, A. M., Balgoma, D., Balboa, M. A. & Balsinde, J. 2012. Dynamics of arachidonic acid mobilization by inflammatory cells. *Biochimica et Biophysica Acta (BBA)-Molecular and Cell Biology of Lipids,* 1821, 249-256.

Aydi, A., Claumann, C. A., Wüst Zibetti, A. & Abderrabba, M. 2016. Differential Scanning Calorimetry Data and Solubility of Rosmarinic Acid in Different Pure Solvents and in Binary Mixtures (Methyl Acetate + Water) and (Ethyl Acetate + Water) from 293.2 to 313.2 K. *Journal of Chemical & Engineering Data,* 61, 3718-3723.

Bais, H. P., Walker, T. S., Schweizer, H. P. & Vivanco, J. M. 2002. Root specific elicitation and antimicrobial activity of rosmarinic acid in hairy root cultures of *Ocimum basilicum. Plant Physiology and Biochemistry,* 40, 983-995.

Bandoniene, D., Murkovic, M. & Venskutonis, P. R. 2005. Determination of rosmarinic acid in sage and borage leaves by high-performance liquid chromatography with different detection methods. *Journal of Chromatographic Science,* 43, 372-376.

Bhatt, R., Singh, D., Prakash, A. & Mishra, N. 2015. Development, characterization and nasal delivery of rosmarinic acid-loaded solid lipid nanoparticles for the effective management of Huntington's disease. *Drug Delivery,* 22, 931-939.

Björkman, M., Klingen, I., Birch, A. N., Bones, A. M., Bruce, T. J., Johansen, T. J., Meadow, R., Mølmann, J., Seljåsen, R. & Smart, L. E. 2011. Phytochemicals of Brassicaceae in plant protection and human health–Influences of climate, environment and agronomic practice. *Phytochemistry,* 72, 538-556.

Boonyarikpunchai, W., Sukrong, S. & Towiwat, P. 2014. Antinociceptive and anti-inflammatory effects of rosmarinic acid isolated from *Thunbergia laurifolia* Lindl. *Pharmacology Biochemistry and Behavior,* 124, 67-73.

Boros, B., Jakabová, S., Dörnyei, Á., Horváth, G., Pluhár, Z., Kilár, F. & Felinger, A. 2010. Determination of polyphenolic compounds by liquid chromatography–mass spectrometry in *Thymus* species. *Journal of Chromatography A,* 1217, 7972-7980.

Brewer, M. 2011. Natural antioxidants: sources, compounds, mechanisms of action, and potential applications. *Comprehensive Reviews in Food Science and Food Safety,* 10, 221-247.

Bulgakov, V. P., Inyushkina, Y. V. & Fedoreyev, S. A. 2012. Rosmarinic acid and its derivatives: biotechnology and applications. *Critical Reviews in Biotechnology,* 32, 203-217.

Calder, P. C. 2011. Fatty acids and inflammation: the cutting edge between food and pharma. *European Journal of Pharmacology,* 668, S50-S58.

Cao, W., Hu, C., Wu, L., Xu, L. & Jiang, W. 2016. Rosmarinic acid inhibits inflammation and angiogenesis of hepatocellular carcinoma by suppression of NF-κB signaling in H22 tumor-bearing mice. *Journal of Pharmacological Sciences,* 132, 131-137.

Carbone, C., Martins-Gomes, C., Caddeo, C., Silva, A. M., Musumeci, T., Pignatello, R., Puglisi, G. & Souto, E. B. 2018. Mediterranean essential oils as precious matrix components and active ingredients of lipid nanoparticles. *Int J Pharm,* 548, 217-226.

Casanova, F., Estevinho, B. N. & Santos, L. 2016. Preliminary studies of rosmarinic acid microencapsulation with chitosan and modified chitosan for topical delivery. *Powder Technology,* 297, 44-49.

Chen, H., Zhang, Q., Wang, X., Yang, J. & Wang, Q. 2011. Qualitative analysis and simultaneous quantification of phenolic compounds in the aerial parts of *Salvia miltiorrhiza* by HPLC-DAD and ESI/MSn. *Phytochemical Analysis,* 22, 247-257.

Chen, S. 2011. Natural products triggering biological targets-a review of the anti-inflammatory phytochemicals targeting the arachidonic acid pathway in allergy asthma and rheumatoid arthritis. *Current Drug Targets,* 12, 288-301.

Chu, X., Ci, X., He, J., Jiang, L., Wei, M., Cao, Q., Guan, M., Xie, X. & Deng, X. 2012. Effects of a natural prolyl oligopeptidase inhibitor, rosmarinic acid, on lipopolysaccharide-induced acute lung injury in mice. *Molecules,* 17, 3586-3598.

Costa, G., Francisco, V., C Lopes, M., T Cruz, M. & T Batista, M. 2012. Intracellular signaling pathways modulated by phenolic compounds: application for new anti-inflammatory drugs discovery. *Current Medicinal Chemistry,* 19, 2876-2900.

Costa, P., Gonçalves, S., Valentão, P., Andrade, P. B., Almeida, C., Nogueira, J. M. & Romano, A. 2013. Metabolic profile and biological activities of *Lavandula pedunculata* subsp. lusitanica (Chaytor) Franco: Studies on the essential oil and polar extracts. *Food Chemistry,* 141, 2501-2506.

Da Silva, S. B., Amorim, M., Fonte, P., Madureira, R., Ferreira, D., Pintado, M. & Sarmento, B. 2015. Natural extracts into chitosan nanocarriers for rosmarinic acid drug delivery. *Pharm Biol,* 53, 642-52.

Da Silva, S. B., Ferreira, D., Pintado, M. & Sarmento, B. 2016. Chitosan-based nanoparticles for rosmarinic acid ocular delivery–*In vitro* tests. *Int J Biol Macromol,* 84, 112-20.

Domitrović, R., Potočnjak, I., Crnčević-Orlić, Ž. & Škoda, M. 2014. Nephroprotective activities of rosmarinic acid against cisplatin-induced kidney injury in mice. *Food and Chemical Toxicology,* 66, 321-328.

Domitrović, R., Škoda, M., Marchesi, V. V., Cvijanović, O., Pugel, E. P. & Štefan, M. B. 2013. Rosmarinic acid ameliorates acute liver damage and fibrogenesis in carbon tetrachloride-intoxicated mice. *Food and Chemical Toxicology,* 51, 370-378.

Dresler, S., Szymczak, G. & Wójcik, M. 2017. Comparison of some secondary metabolite content in the seventeen species of the Boraginaceae family. *Pharmaceutical Biology,* 55, 691-695.

Dubois, M., Bailly, F., Mbemba, G., Mouscadet, J.-F., Debyser, Z., Witvrouw, M. & Cotelle, P. 2008. Reaction of rosmarinic acid with nitrite ions in acidic conditions: discovery of nitro-and dinitrorosmarinic acids as new anti-HIV-1 agents. *Journal of Medicinal Chemistry,* 51, 2575-2579.

El-Hattab, A. W., Emrick, L. T., Craigen, W. J. & Scaglia, F. 2012. Citrulline and arginine utility in treating nitric oxide deficiency in mitochondrial disorders. *Molecular Genetics and Metabolism,* 107, 247-252.

Embuscado, M. E. 2015. Spices and herbs: Natural sources of antioxidants–a mini review. *Journal of Functional Foods,* 18, 811-819.

Fadel, O., El Kirat, K. & Morandat, S. 2011. The natural antioxidant rosmarinic acid spontaneously penetrates membranes to inhibit lipid peroxidation in situ. *Biochimica et Biophysica Acta (BBA)-Biomembranes,* 1808, 2973-2980.

Fan, H. Y., Fu, F. H., Yang, M. Y., Xu, H., Zhang, A. H. & Liu, K. 2010. Antiplatelet and antithrombotic activities of salvianolic acid A. *Thrombosis Research,* 126, e17-e22.

Fernando, P. M. D. J., Piao, M. J., Kang, K. A., Ryu, Y. S., Hewage, S. R. K. M., Chae, S. W. & Hyun, J. W. 2016. Rosmarinic acid attenuates cell damage against UVB radiation-induced oxidative stress via enhancing antioxidant effects in human HaCaT cells. *Biomolecules & Therapeutics,* 24, 75.

Fujimoto, A., Shingai, Y., Nakamura, M., Maekawa, T., Sone, Y. & Masuda, T. 2010. A novel ring-expanded product with enhanced tyrosinase inhibitory activity from classical Fe-catalyzed oxidation of rosmarinic acid, a potent antioxidative *Lamiaceae* polyphenol. *Bioorganic & Medicinal Chemistry Letters,* 20, 7393-7396.

Furtado, M. A., De Almeida, L. C. F., Furtado, R. A., Cunha, W. R. & Tavares, D. C. 2008. Antimutagenicity of rosmarinic acid in Swiss mice evaluated by the micronucleus assay. *Mutation Research/Genetic Toxicology and Environmental Mutagenesis,* 657, 150-154.

Gamaro, G. D., Suyenaga, E., Borsoi, M., Lermen, J., Pereira, P. & Ardenghi, P. 2011. Effect of rosmarinic and caffeic acids on inflammatory and nociception process in rats. *ISRN Pharmacology,* 2011.

Gülçin, İ., Scozzafava, A., Supuran, C. T., Koksal, Z., Turkan, F., Çetinkaya, S., Bingöl, Z., Huyut, Z. & Alwasel, S. H. 2016. Rosmarinic acid inhibits some metabolic enzymes including glutathione S-transferase, lactoperoxidase, acetylcholinesterase, butyrylcholinesterase and carbonic anhydrase isoenzymes. *Journal of Enzyme Inhibition and Medicinal Chemistry,* 31, 1698-1702.

Han, S., Yang, S., Cai, Z., Pan, D., Li, Z., Huang, Z., Zhang, P., Zhu, H., Lei, L. & Wang, W. 2015. Anti-Warburg effect of rosmarinic acid via miR-155 in gastric cancer cells. *Drug Design, Development and Therapy,* 9, 2695.

Harbourne, N., Marete, E., Jacquier, J. C. & O'riordan, D. 2013. Conventional extraction techniques for phytochemicals. *Handbook of Plant Food Phytochemicals: Sources, Stability and Extraction*, 397-411.

Jabri-Karoui, I., Bettaieb, I., Msaada, K., Hammami, M. & Marzouk, B. 2012. Research on the phenolic compounds and antioxidant activities of Tunisian *Thymus capitatus. Journal of Functional Foods,* 4, 661-669.

Kelsey, N. A., Wilkins, H. M. & Linseman, D. A. 2010. Nutraceutical antioxidants as novel neuroprotective agents. *Molecules,* 15, 7792-7814.

Khattak, S., Wahid, F., Liu, L. P., Jia, S. R., Chu, L. Q., Xie, Y. Y., Li, Z. X. & Zhong, C. 2019. Applications of cellulose and chitin/chitosan derivatives and composites as antibacterial materials: current state and perspectives. *Appl Microbiol Biotechnol.* 103:1989.

Kim, G.-D., Park, Y. S., Jin, Y.-H. & Park, C.-S. 2015. Production and applications of rosmarinic acid and structurally related compounds. *Applied Microbiology and Biotechnology,* 99, 2083-2092.

Kuo, Y. C. & Rajesh, R. 2017. Targeted delivery of rosmarinic acid across the blood-brain barrier for neuronal rescue using polyacrylamide-chitosan-poly(lactide-co-glycolide) nanoparticles with surface cross-reacting material 197 and apolipoprotein E. *Int J Pharm,* 528, 228-241.

Kuo, Y. C. & Tsai, H. C. 2018. Rosmarinic acid- and curcumin-loaded polyacrylamide-cardiolipin-poly(lactide-co-glycolide) nanoparticles with conjugated 83-14 monoclonal antibody to protect beta-amyloid-insulted neurons. *Mater Sci Eng C Mater Biol Appl,* 91, 445-457.

Lee, J., Kim, Y. S. & Park, D. 2007. Rosmarinic acid induces melanogenesis through protein kinase A activation signaling. *Biochemical Pharmacology,* 74, 960-968.

Li, G.-S., Jiang, W.-L., Tian, J.-W., Qu, G.-W., Zhu, H.-B. & Fu, F.-H. 2010. In vitro and in vivo antifibrotic effects of rosmarinic acid on experimental liver fibrosis. *Phytomedicine,* 17, 282-288.

Lim, S. H. & Park, Y. 2018. Green Synthesis, Characterization and Catalytic Activity of Gold Nanoparticles Prepared Using Rosmarinic Acid. *J Nanosci Nanotechnol,* 18, 659-667.

Luan, H., Kan, Z., Xu, Y., Lv, C. & Jiang, W. 2013. Rosmarinic acid protects against experimental diabetes with cerebral ischemia: relation to inflammation response. *Journal of Neuroinflammation,* 10, 810.

Madureira, A. R., Campos, D. A., Oliveira, A., Sarmento, B., Pintado, M. M. & Gomes, A. M. 2016a. Insights into the protective role of solid lipid nanoparticles on rosmarinic acid bioactivity during exposure to simulated gastrointestinal conditions. *Colloids and Surfaces B: Biointerfaces,* 139, 277-284.

Madureira, A. R., Nunes, S., Campos, D. A., Fernandes, J. C., Marques, C., Zuzarte, M., Gullon, B., Rodriguez-Alcala, L. M., Calhau, C., Sarmento, B., Gomes, A. M., Pintado, M. M. & Reis, F. 2016b. Safety profile of solid lipid nanoparticles loaded with rosmarinic acid for oral use: in vitro and animal approaches. *Int J Nanomedicine,* 11, 3621-40.

Martins-Gomes, C., Taghouti, M., Schäfer, J., Bunzel, M., Silva, A. M. & Nunes, F. M. 2018. Chemical characterization and bioactive properties of decoctions and hydroethanolic extracts of *Thymus carnosus* Boiss. *Journal of Functional Foods,* 43, 154-164.

Martins, N., Barros, L., Santos-Buelga, C., Silva, S., Henriques, M. & Ferreira, I. C. 2015. Decoction, infusion and hydroalcoholic extract of cultivated thyme: Antioxidant and antibacterial activities, and phenolic characterisation. *Food Chemistry,* 167, 131-137.

Moon, D.-O., Kim, M.-O., Lee, J.-D., Choi, Y. H. & Kim, G.-Y. 2010. Rosmarinic acid sensitizes cell death through suppression of TNF-α-induced NF-κB activation and ROS generation in human leukemia U937 cells. *Cancer Letters,* 288, 183-191.

Moradabadi, L., Kouhsari, S. M. & Sani, M. F. 2013. Hypoglycemic Effects of Three Medicinal plants in experimental diabetes: Inhibition of rat intestinal α-glucosidase and enhanced pancreatic insulin and cardiac Glut-4 mRNAs expression. *Iranian Journal of Pharmaceutical Research: IJPR,* 12, 387.

Mueller, M., Hobiger, S. & Jungbauer, A. 2010. Anti-inflammatory activity of extracts from fruits, herbs and spices. *Food Chemistry,* 122, 987-996.

Muñoz-Muñoz, J., Garcia-Molina, F., Ros, E., Tudela, J., García-Canovas, F. & Rodriguez-Lopez, J. 2013. Prooxidant and antioxidant activities of rosmarinic acid. *Journal of Food Biochemistry,* 37, 396-408.

Naimi, M., Vlavcheski, F., Shamshoum, H. & Tsiani, E. 2017. Rosemary extract as a potential anti-hyperglycemic agent: current evidence and future perspectives. *Nutrients,* 9, 968.

Navarro-González, J. F., Mora-Fernández, C., De Fuentes, M. M. & García-Pérez, J. 2011. Inflammatory molecules and pathways in the pathogenesis of diabetic nephropathy. *Nature Reviews Nephrology,* 7, 327.

Oh, H.-A., Park, C.-S., Ahn, H.-J., Park, Y. S. & Kim, H.-M. 2011a. Effect of *Perilla frutescens* var. acuta Kudo and rosmarinic acid on allergic inflammatory reactions. *Experimental Biology and Medicine,* 236, 99-106.

Oh, K.-S., Oh, B. K., Mun, J., Seo, H. W. & Lee, B. H. 2011b. Salvianolic acid A suppress lipopolysaccharide-induced NF-κB signaling pathway by targeting IKKβ. *International Immunopharmacology,* 11, 1901-1906.

Pei, R., Si, T., Lu, Y., Zhou, J. X. & Jiang, L. 2018. Salvianolic acid A, a novel PI3K/Akt inhibitor, induces cell apoptosis and suppresses tumor growth in acute myeloid leukemia. *Leukemia & Lymphoma,* 59, 1959-1967.

Pereira, I., Zielinska, A., Ferreira, N. R., Silva, A. M. & Souto, E. B. 2018a. Optimization of linalool-loaded solid lipid nanoparticles using experimental factorial design and long-term stability studies with a new centrifugal sedimentation method. *Int J Pharm,* 549, 261-270.

Pereira, O., Catarino, M., Afonso, A., Silva, A. & Cardoso, S. 2018b. *Salvia elegans*, *Salvia greggii* and *Salvia officinalis* Decoctions: Antioxidant Activities and Inhibition of Carbohydrate and Lipid Metabolic Enzymes. *Molecules,* 23, 3169.

Petersen, M. 2013. Rosmarinic acid: new aspects. *Phytochemistry Reviews,* 12, 207-227.

Petersen, M., Häusler, E., Karwatzki, B. & Meinhard, J. 1993. Proposed biosynthetic pathway for rosmarinic acid in cell cultures of Coleus blumei Benth. *Planta,* 189, 10-14.

Petersen, M. & Simmonds, M. S. 2003a. Rosmarinic acid. *Phytochemistry,* 62, 121-125.

Petersen, M. & Simmonds, M. S. J. 2003b. Rosmarinic acid. *Phytochemistry,* 62, 121-125.

Qiao, S., Li, W., Tsubouchi, R., Haneda, M., Murakami, K., Takeuchi, F., Nisimoto, Y. & Yoshino, M. 2005. Rosmarinic acid inhibits the formation of reactive oxygen and nitrogen species in RAW264. 7 macrophages. *Free Radical Research,* 39, 995-1003.

Rahbardar, M. G., Amin, B., Mehri, S., Mirnajafi-Zadeh, S. J. & Hosseinzadeh, H. 2017. Anti-inflammatory effects of ethanolic extract of *Rosmarinus officinalis* L. and rosmarinic acid in a rat model of neuropathic pain. *Biomedicine & Pharmacotherapy,* 86, 441-449.

Rocha, J., Eduardo-Figueira, M., Barateiro, A., Fernandes, A., Brites, D., Bronze, R., Duarte, C. M., Serra, A. T., Pinto, R. & Freitas, M. 2015. Anti-inflammatory effect of rosmarinic acid and an extract of *Rosmarinus officinalis* in rat models of local and systemic inflammation. *Basic & Clinical Pharmacology & Toxicology,* 116, 398-413.

Runtuwene, J., Cheng, K.-C., Asakawa, A., Amitani, H., Amitani, M., Morinaga, A., Takimoto, Y., Kairupan, B. H. R. & Inui, A. 2016. Rosmarinic acid ameliorates hyperglycemia and insulin sensitivity in diabetic rats, potentially by modulating the expression of PEPCK and GLUT4. *Drug Design, Development and Therapy,* 10, 2193.

Santangelo, C., Varì, R., Scazzocchio, B., Di Benedetto, R., Filesi, C. & Masella, R. 2007. Polyphenols, intracellular signalling and inflammation. *Annali-Istituto Superiore di Sanita,* 43, 394.

Scarpati, M. L. & Oriente, G. 1958. Isolamento e costituzione dell'acido rosmarinico (dal *Rosmarinus off.*). *Ric Sci,* 28, 2329-2333.

Sevgi, K., Tepe, B. & Sarikurkcu, C. 2015. Antioxidant and DNA damage protection potentials of selected phenolic acids. *Food and Chemical Toxicology,* 77, 12-21.

Shariatinia, Z. 2019. Pharmaceutical applications of chitosan. *Adv Colloid Interface Sci,* 263, 131-194.

Sharmila, R. & Manoharan, S. 2012. Anti-tumor activity of rosmarinic acid in 7, 12-dimethylbenz (a) anthracene (DMBA) induced skin carcinogenesis in Swiss albino mice. *Indian Journal of Experimental Biology* 50:187-194.

Shen, D., Pan, M.-H., Wu, Q.-L., Park, C.-H., Juliani, H. R., Ho, C.-T. & Simon, J. E. 2010. LC-MS method for the simultaneous quantitation of the anti-inflammatory constituents in oregano (*Origanum* species). *Journal of Agricultural and Food Chemistry,* 58, 7119-7125.

Sotnikova, R., Okruhlicova, L., Vlkovicova, J., Navarova, J., Gajdacova, B., Pivackova, L., Fialova, S. & Krenek, P. 2013. Rosmarinic acid administration attenuates diabetes-induced vascular dysfunction of the rat aorta. *Journal of Pharmacy and Pharmacology,* 65, 713-723.

Swarup, V., Ghosh, J., Ghosh, S., Saxena, A. & Basu, A. 2007. Antiviral and anti-inflammatory effects of rosmarinic acid in an experimental murine model of Japanese encephalitis. *Antimicrobial Agents and Chemotherapy,* 51, 3367-3370.

Taghouti, M., Martins-Gomes, C., Schäfer, J., Félix, L. M., Santos, J. A., Bunzel, M., Nunes, F. M. & SilvA, A. M. 2018. *Thymus pulegioides* L. as a rich source of antioxidant, anti-proliferative and neuroprotective phenolic compounds. *Food & Function.*

Terpinc, P. & Abramovič, H. 2010. A kinetic approach for evaluation of the antioxidant activity of selected phenolic acids. *Food Chemistry,* 121, 366-371.

Topal, M. & Gulçin, İ. 2014. Rosmarinic acid: a potent carbonic anhydrase isoenzymes inhibitor. *Turkish Journal of Chemistry,* 38, 894-902.

Vostálová, J., Zdařilová, A. & Svobodová, A. 2010. *Prunella vulgaris* extract and rosmarinic acid prevent UVB-induced DNA damage and oxidative stress in HaCaT keratinocytes. *Archives of Dermatological Research,* 302, 171-181.

Wang, H., Provan, G. J. & Helliwell, K. 2004. Determination of rosmarinic acid and caffeic acid in aromatic herbs by HPLC. *Food Chemistry,* 87, 307-311.

Wang, J., Pan, X., Han, Y., Guo, D., Guo, Q. & Li, R. 2012. Rosmarinic acid from eelgrass shows nematicidal and antibacterial activities against pine wood nematode and its carrying bacteria. *Marine Drugs,* 10, 2729-2740.

Wang, S.-X., Hu, L.-M., Gao, X.-M., Guo, H. & Fan, G.-W. 2010. Anti-inflammatory activity of salvianolic acid B in microglia contributes to its neuroprotective effect. *Neurochemical Research,* 35, 1029-1037.

Wu, C.-F., Hong, C., Klauck, S. M., Lin, Y.-L. & Efferth, T. 2015. Molecular mechanisms of rosmarinic acid from *Salvia miltiorrhiza* in acute lymphoblastic leukemia cells. *Journal of Ethnopharmacology,* 176, 55-68.

Xu, Y., Xu, G., Liu, L., Xu, D. & Liu, J. 2010. Anti-invasion effect of rosmarinic acid via the extracellular signal-regulated kinase and oxidation-reduction pathway in Ls174-T cells. *Journal of Cellular Biochemistry,* 111, 370-379.

Yan, Q., Shi, M., Ng, J. & Wu, J. Y. 2006. Elicitor-induced rosmarinic acid accumulation and secondary metabolism enzyme activities in Salvia miltiorrhiza hairy roots. *Plant Science,* 170, 853-858.

Yang, E. J., Ku, S. K., Lee, W., Lee, S., Lee, T., Song, K. S. & Bae, J. S. 2013. Barrier protective effects of rosmarinic acid on HMGB1-induced inflammatory responses *in vitro* and *in vivo*. *Journal of Cellular Physiology,* 228, 975-982.

Yoo, H. G., Lee, B. H., Kim, W., Lee, J. S., Kim, G. H., Chun, O. K., Koo, S. I. & Kim, D.-O. 2014. *Lithospermum erythrorhizon* extract protects keratinocytes and fibroblasts against oxidative stress. *Journal of Medicinal Food,* 17, 1189-1196.

Zhang, Y., Chen, X., Yang, L., Zu, Y. & Lu, Q. 2015. Effects of rosmarinic acid on liver and kidney antioxidant enzymes, lipid peroxidation and tissue ultrastructure in aging mice. *Food & Function,* 6, 927-931.

Zhu, F. 2015. Interactions between starch and phenolic compound. *Trends in Food Science & Technology,* 43, 129-143.

Zielinska, A., Martins-Gomes, C., Ferreira, N. R., Silva, A. M., Nowak, I. & Souto, E. B. 2018. Anti-inflammatory and anti-cancer activity of citral: Optimization of citral-loaded solid lipid nanoparticles (SLN) using experimental factorial design and LUMiSizer(R). *Int J Pharm,* 553, 428-440.

Zimmermann, B. F., Walch, S. G., Tinzoh, L. N., Stühlinger, W. & Lachenmeier, D. W. 2011. Rapid UHPLC determination of polyphenols in aqueous infusions of *Salvia officinalis* L. (sage tea). *Journal of Chromatography B,* 879, 2459-2464.

In: Phytochemicals
Editor: Iman Ryan
ISBN: 978-1-53615-478-8

Chapter 5

BRAZILIAN *SMILAX* SPECIES AS PROMISING SOURCES OF BIOACTIVE COMPOUNDS

Paula Avelar Amado, Thaís Paula Rodrigues Gonçalves, Iara Cotta Mendes, Ana Hortência Fonsêca Castro and Luciana Alves Rodrigues dos Santos Lima
Campus Centro-Oeste Dona Lindu,
Federal University of São João Del Rei, Divinópolis,
Minas Gerais, Brazil

ABSTRACT

Brazil stands out as one of the most biodiverse countries on the planet. It is estimated that Brazilian ecosystems house about 20% of all biological species in the world. The medicinal plants of Brazil have stood out for their importance in popular use. It is widely documented and proven that medicinal plants are a source of unpublished molecules. The family *Smilacaceae* belongs to the group of monocotyledons, possessing 310 species distributed in temperate, tropical and subtropical regions. The family consists of dioecious species, vines or herbaceous vines, and may occur as sub-shrubs or shrubs, with simple, alternating, petiolate leaves with reticulated veins. *Smilax* is the largest genus in the *Smilacaceae* family and is represented in Brazil by 33 species widely distributed throughout the country, but mainly in the Cerrado. Plants of this genus are known as "sarsaparilla" and "japecanga" and are used in folk medicine as a tonic, antirheumatic, anti-syphilitic, antihypertensive and diuretic and for sweating, arthritis and skin conditions. The genus *Smilax* has proved to be a source of secondary metabolites that have several types of biological activity. Phytochemical investigations have shown the presence of polar compounds such as: saponins and a complex mixture of phenolic compounds such as flavonoids, caffeic acid, *p*-coumaric acid, ferulic acid and *trans*-cinnamic acid in the species present in Brazil. Regarding the importance of studies of species of the genus *Smilax*, few studies are available in the literature. The biological activity already reported for Brazilian species of this genus is antioxidant, antifungal, antibacterial, antinociceptive, anti-inflammatory and allelopathic in nature. The objective of this review is to report aspects of the chemical composition and biological activity of the species of the genus *Smilax* found in Brazil.

Keywords: *Smilacaceae*, *Smilax*, chemical composition, biological activity

1. THE FAMILY *SMILACACEAE* AND THE GENUS *SMILAX*

Brazil is considered one of the most biodiverse countries in the world, since it is estimated that more than 20% of all biological species in the world are found there [1]. However, only 8% of the plant species of the Brazilian flora have been studied in relation to their biological potential, and only 1,100 species have been evaluated for their medicinal properties [2].

Among the five areas abundant in native plants in Brazil, the Cerrado biome stands out. This biome is characterized as a great holder of biological diversity, being the savanna formation with the greatest plant diversity in the world [3]. Among the various species in popular use present in the Cerrado are species of the family *Smilacaceae*.

The family *Smilacaceae* belongs to the order Liliales and to the group of monocotyledons, constituted by five genera, *Anikenton*, *Coprosmanthus*, *Heterosmilax*, *Nemexia* and *Smilax* [4], with 310 species distributed in temperate, tropical and subtropical regions of the whole planet [5, 6]. The family is composed of herbaceous plants, vines or herbaceous vines, usually little erect. They are also described as shrubs or sub-shrubs. The leaves are simple, alternate, petiolate and with reticulated veins. The stem is aerial and, in general, aculeous [7].

Smilax is the largest genus of the family *Smilacaceae* [8]. Species of this genus are popularly known as japecanga, japonanga, japuanga, japonica, jupicanga, inhapecanga, inhapican, nhupicanga, salsaparrilha-de-minas, salsaparrilha-das-boticas, sarza, zarza and salsa-americana [9, 10]. According to Silva Júnior [11], the different species of *Smilax* present as prostrate, perennial, rhizomatous and almost always with thorns.

The species of the genus *Smilax* occurring in Brazil are woody lianas, small or large, shrubs or sub-bushes; they have aculeous stems and branches. Their leaves are simple, alternate and glabrous, with a well-differentiated blade, petiole and hem. The petiole, in addition to presenting a pair of tendrils, may also present twists that determine leaf laminar position [12].

In Brazil, this genus is represented by 33 species distributed in several ecosystems, mainly in the Cerrado, Amazon Forest, Caatinga and Riparian Forests [10, 12].

The species of *Smilax* found in Brazil are: *S. brasiliensis*, *S. campestris*, *S. cissoides*, *S. cognata*, *S. elastica*, *S. fluminensis*, *S. goyazana*, *S. havanensis*, *S. hilariana*, *S. irrorata*, *S. japicanga*, *S. krukovii*, *S. lappacea*, *S. larvata*, *S. longifolia*, *S. lutescens*, *S. minarum*, *S. muscosa*, *S. oblongifolia*, *S. pilosa*, *S. polyantha*, *S. quinquenervia*, *S. remotinervis*, *S. rufescens*, *S. salicifolia*, *S. santaremensis*, *S. spicata*, *S. spruceana*, *S. staminea*, *S. stenophylla*, *S. subsessiliflora*, *S. syphilitica* and *S. verrucosa* [10]. However, of the 33 species found in Brazil, only 14 are Brazilian and only seven have records of their medicinal properties in ethnobotanical surveys [12].

Species of the genus *Smilax* have been extensively studied because of their use in folk medicine throughout the world. There is a great variety of reports of the use of the leaves, fruits and roots of the species of this genus. *S. brasiliensis* is used in popular medicine for rheumatism, against syphilis and as a capillary tonic [13, 14], and as a therapeutic agent in immune-inflammatory diseases such as rheumatoid arthritis [15]. *S. campestris* is commonly used in folk medicine for the treatment of syphilis, rheumatism and some skin diseases, and as a diaphoretic and diuretic [16]. *S. officinalis* is used as an antirheumatic, against syphilis and as a diuretic [17]. *S. larvata* has been used in folk medicine as an anti-inflammatory agent [18]. Studies attribute to *S. quinquenervia* analgesic [19] as well as antioxidant properties [20]. To *S. goyazana* are attributed depurative, antirheumatic and diuretic properties and it is used for liver problems [21].

2. Phytochemical Study of the Genus *Smilax*

Research on the phytochemical composition of *Smilax* species is found in the literature. The genus *Smilax* has been shown to be a source of secondary metabolites, especially polar compounds, such as saponins, phytosterols, triterpenoids and a complex mixture of phenolic compounds such as flavonoids, phenolic acids and tannins, which present several types of biological activity [22-24].

A study with the aerial parts of *S. larvata* [25] showed the presence of two diglycosylated flavonoids, drabanemoroside and kaempferol 3-*O*-α-L-rhamnopyranosyl (1→2)-α-L-rhamnopyranoside, as well as kaempferol, *p*-coumaric acid and *p*-hydroxybenzoic acid.

The extracts of roots, rhizophores and leaves of *S. fluminensis* [23] revealed the presence of chlorogenic acid, caffeic acid, rutin, *p*-coumaric acid and *trans*-cinnamic acid.

The chemical profile of some *Smilax* species native to Brazil was determined in the Martins [26] study, which evaluated *S. brasiliensis*, *S. campestris*, *S. cissoides* and *S. polyantha* species. The methanol extracts of the root, rhizophores and leaves of *S. brasiliensis*, *S. campestris*, *S. cissoides* and *S. polyantha* were analyzed by high-performance liquid chromatography and revealed the presence of several substances. However, only the peaks corresponding to chlorogenic acid, caffeic acid, rutin, *p*-coumaric acid, ferulic acid and *trans*-cinnamic acid were identified.

Martins [26] also verified that chlorogenic, *p*-coumaric and ferulic acids are common in the secondary metabolism of *S. brasiliensis*, *S. campestris*, *S. cissoides*, *S. fluminensis* and *S. polyantha* species.

The metabolites identified in the extract and fractions from *S. brasiliensis* [27] analyzed by Liquid Chromatography coupled to a Diode Array Detector and Mass Spectrometer (LC-DAD-MS) include chlorogenic acid, flavonoids, phenylpropanoids and others. The constituents identified were: quinic acid, gallocatechin, 3-*O*-E-caffeoyl quinic acid, *O*-feruloyl quinic acid, *O*-caffeoyl shikimic acid, ferulic acid, 3-*O*-β-galactopyranosyl quercetin and 3-*O*-β-glycopyranosyl quercetin, quercetin, naringenin and kaempferol (Figure 1). Analysis by Thin Layer Chromatography (TLC) compared to standards suggested the presence of rutin in the ethanol extract and in the ethyl acetate and hydroethanol fractions of *S. brasiliensis* leaves. In addition, chlorogenic acid was present in the ethanol extract and in the dichloromethane, ethyl acetate and hydroethanol fractions [28].

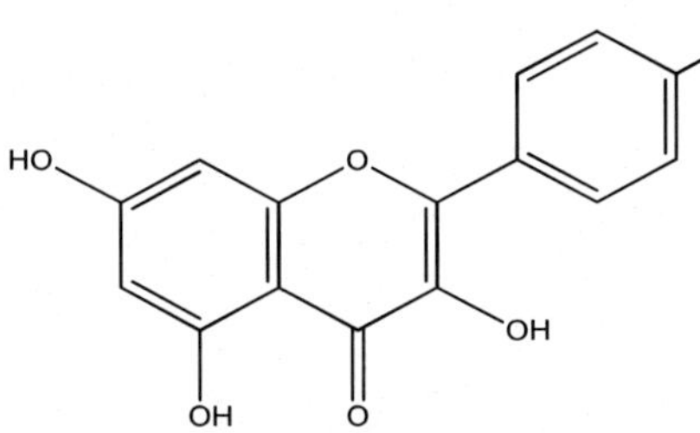

Kaempferol

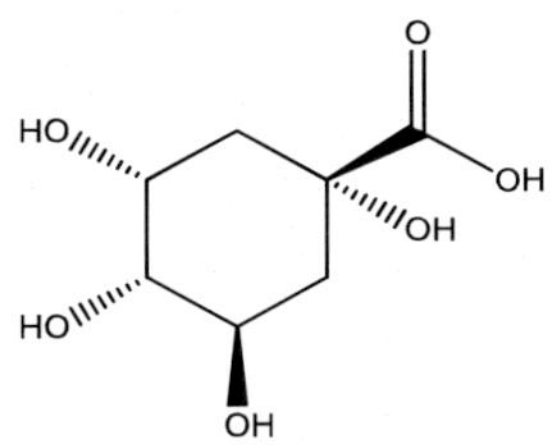

Quinic Acid

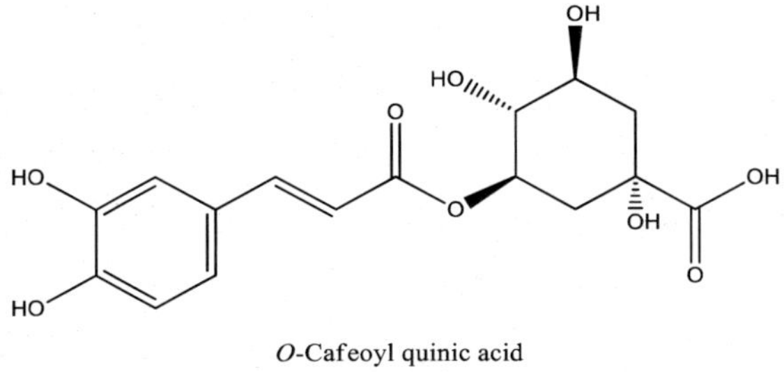

O-Cafeoyl quinic acid

Figure 1. (Continued).

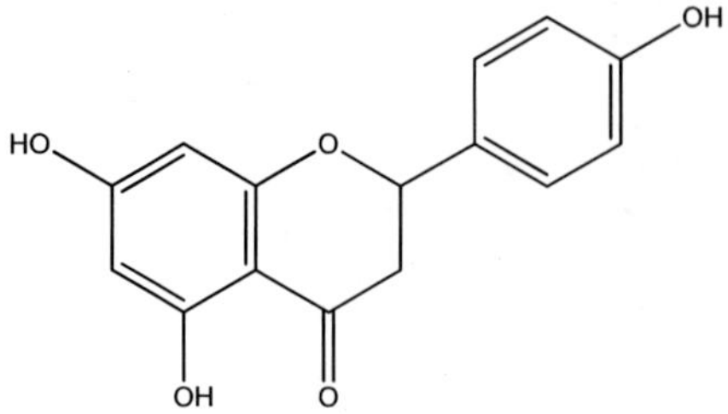

Naringenin

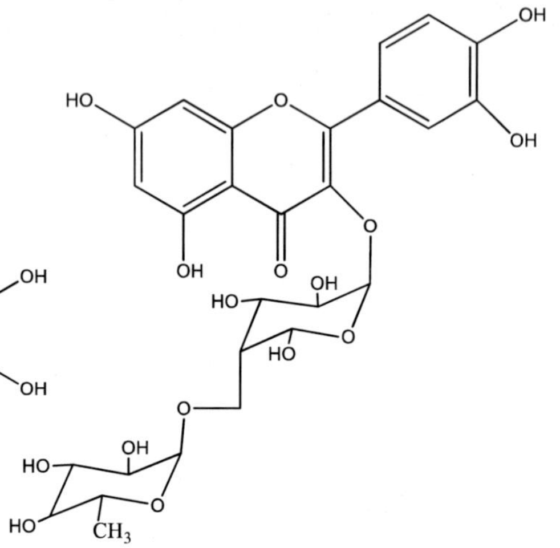

Rutin

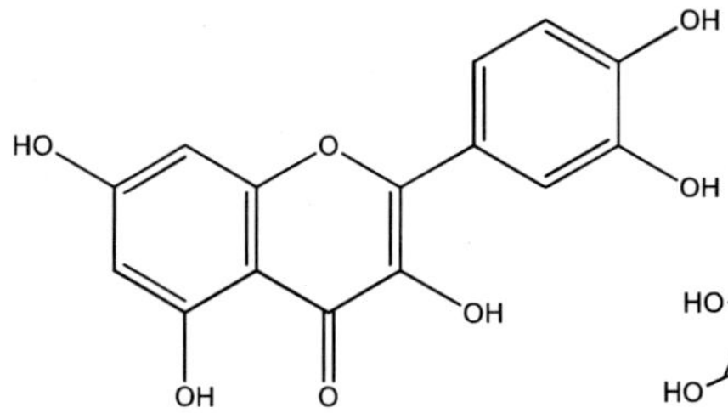

Quercetin

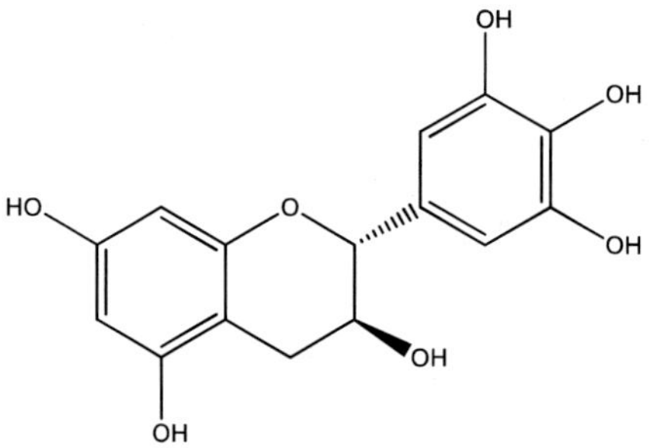

Gallocatechin

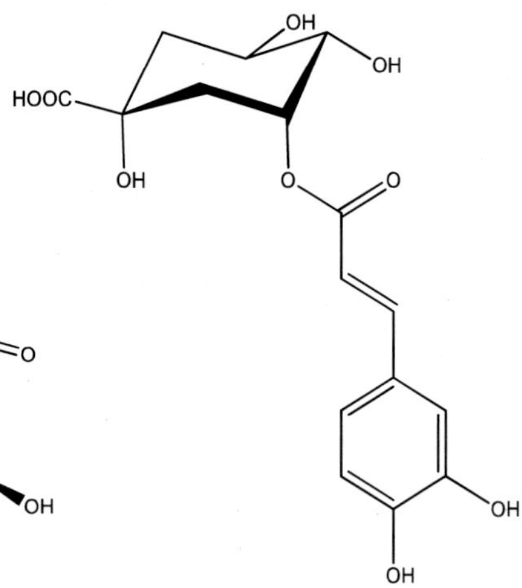

3-*O*-E-Cafeoyl Quinic Acid

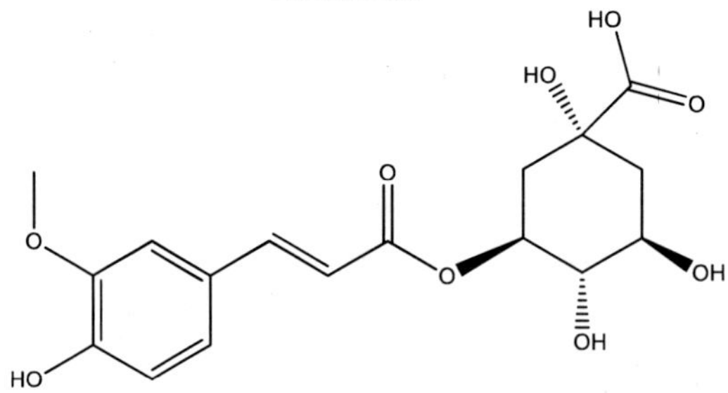

O-Feruloyl Quinic Acid

Figure 1. (Continued).

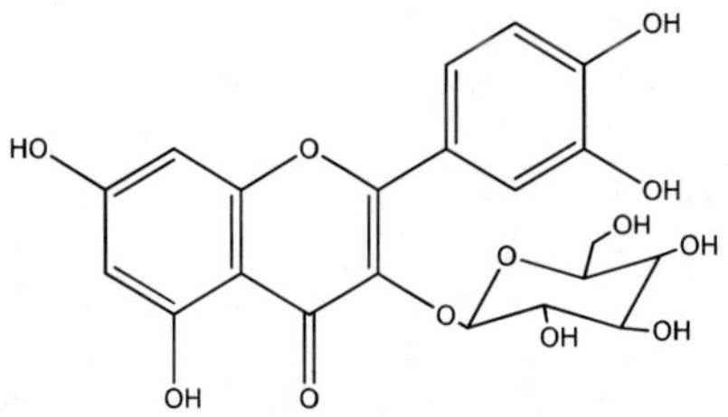

3-*O*-β-Glicopyranosyl quercetin

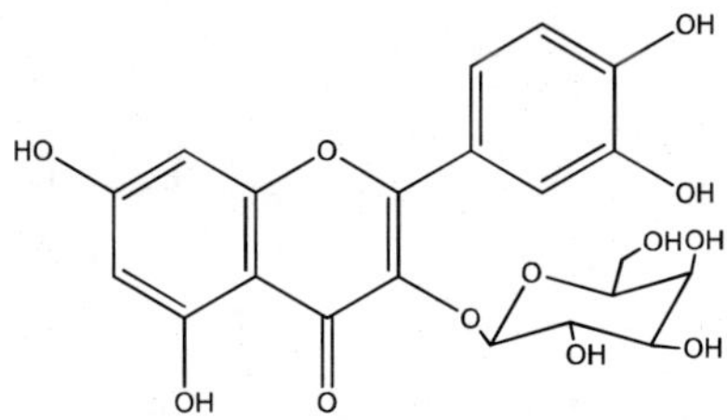

3-*O*-β-Galactopyranosyl quercetin

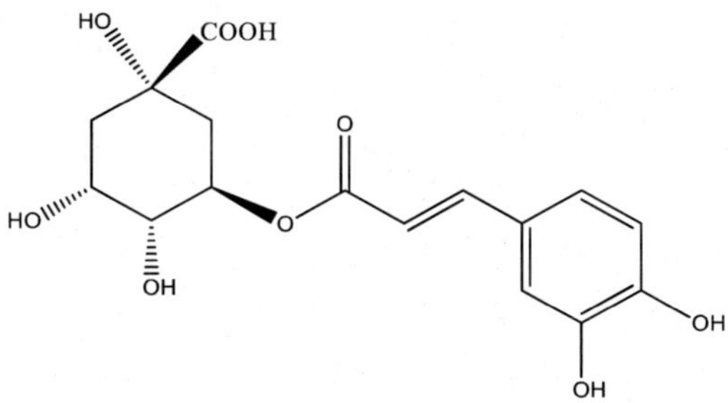

Chlorogenic Acid

Figure 1. Structures of some substances identified in species of the genus *Smilax* native to Brazil.

Extracts obtained from *S. brasiliensis* leaves were investigated by Gas Chromatography-Mass Spectrometry (GC-MS) in their methyl ester form, where it was possible to identify the presence of methyl laurate, methyl myristate, methyl palmitate, methyl palmitoleate, methyl margarate, methyl stearate, methyl oleate, methyl linoleate, methyl linolenate, methyl eicosanoate, methyl docasanoate, methyl tetracosanoate and methyl octacosanoate [29].

3. BIOLOGICAL ACTIVITY OF THE GENUS *SMILAX*

Due to their intense use in traditional medicine, investigations into the possible biological activity of *Smilax* species have been carried out.

3.1. Antioxidant Activity

The 1,1-diphenyl-2-picrylhydrazyl (DPPH) method was used to estimate the antioxidant activity of aerial parts of *S. campestris*. 2,6-Di-tert-butyl-4-methylphenol (BHT) and ascorbic acid were used as standards. The butanol fraction and ethanol extract showed higher antioxidant activity in comparison with other fractions; in addition, these samples exhibited high levels of phenolic compounds that, after analysis by high-performance liquid chromatography, indicated the presence of quercetin derivatives, which concludes the antioxidant activity observed [30].

Using the DPPH assay and BHT as standards, it was concluded that the extract of the aerial parts of *S. larvata* has moderate antioxidant activity with an IC_{50} (inhibition concentration) value of 58.17 μg/mL. This antioxidant activity can be attributed to the amount of phenolic compounds that the extract presented [25].

The DPPH assay was also used to evaluate the antioxidant activity of *S. brasiliensis* leaves. In the study by Fonseca et al. [28], BHT and ascorbic acid were used as standards, the extract and fractions being more effective in eliminating the DPPH radical than the BHT standard, with IC_{50} = 2.73 - 11.75 μg/mL. Compared with the ascorbic acid standard, the extract and fractions were less effective. In the study by Amado et al. [29] carried out by the DPPH assay using BHT as standard, petroleum ether and methanol extracts and fatty acids and methyl esters of fatty acids obtained from leaves of *S. brasiliensis* were evaluated. All of the substances tested showed more potent antioxidant activity with lower IC_{50} values (0.240 - 0.875 μg/mL) and showed higher activity than BHT (IC_{50} = 16.36 μg/mL).

3.2. Antifungal and Antibacterial Activity

The antifungal activity of ethanol extract and fractions of aerial parts of *S. campestris* was evaluated against seven yeast strains of clinical interest. The ethyl acetate fraction exhibited the highest activity against *Candida krusei*, with a minimal inhibitory concentration (MIC) of 62.5

μg/mL and was also active against *Cryptococcus gattii*, with an MIC of 1000 μg/mL. The hexane fraction exhibited activity against *Candida glabrata* and *Candida krusei*, with MIC of 125 and 500 μg/mL, respectively [28]. The ethanol extract was active against *Candida krusei*, *Candida parapsilosis* and *Cryptococcus gattii* with an MIC of 1000 μg/mL, and against *Candida albicans*, *Candida glabrata* and *Candida tropicalis* with an MIC of 2000 μg/mL. The dichloromethane fraction displayed activity against *Candida krusei* and *Candida tropicalis*, with an MIC of 2000 μg/mL. Meanwhile, the butanol and hydroalcoholic fractions were inactive against all the yeast strains studied [28].

Aqueous and methanol extracts of the leaves, rhizomes and roots of *S. campestris* were used to evaluate antifungal activity against *Candida albicans* and *Aspergillus niger* [31]. Methanol extracts obtained from the leaves and rhizomes of *S. campestris* presented activity against *Candida albicans*; however, methanol extracts of the rhizomes were more potent than that of the leaves [31].

The antifungal activity of the aqueous, methanol and dichloromethane extracts of the leaves, roots and branches of *S. brasiliensis* was tested under *Candida albicans*, *Candida glabrata* and *Candida parapsilosis*. All samples exhibited activity, except for the dichloromethane extract from the branches that was inactive under the strains tested [32].

Ethanol extract and hexane, chloroform, ethyl acetate and hydroethanol fractions of aerial parts of *S. larvata* were tested in strains of *Staphylococcus aureus*, *Staphylococcus epidermidis*, *Escherichia coli*, *Pseudomonas aeruginosa*, *Agrobacterium tumefaciens* and *Candida albicans*; the ethanol extract was the only one with activity against the strain *Candida albicans*. All other extracts were inactive for the strains [18].

The antibacterial and antifungal activity of *S. brasiliensis* was also reported in the study by Silva et al. [33]. The antibacterial activity of the hydroethanol extract was tested under *Escherichia coli*, *Klebsiella pneumoniae*, *Pseudomonas aeruginosa*, *Salmonella typhimurium*, *Shigella flexneri*, *Helicobacter pylori*, *Enterococcus faecalis*, *Staphylococcus aureus*, *Staphylococcus epidermidis*, *Streptococcus pyogenes* and *Bacillus subtilis*. However, the extract presented weak activity. Weak activity was also demonstrated when the extract was tested under *Candida albicans*, fluconazole-resistant *Candida albicans*, *Candida parapsilosis*, *Candida tropicalis*, *Cryptococcus neoformans*, *Candida glabrata*, *Aspergillus fumigatus*, *Aspergillus niger*, *Aspergillus parasiticus*, *Aspergillus terreus*, *Penicillium verrucosum*, *Trichophyton mentagrophytes*, *Trichophyton rubrum* and *Microsporum gypseum* [33].

An extract of *S. rufescens* was evaluated for its anthelmintic activity under *Staphylococcus aureus*, *Escherichia coli*, *Enterococcus faecalis* and *Pseudomonas aeruginosa*. The extract was active for *Enterococcus faecalis* [34].

3.3. Anti-Inflammatory, Antinociceptive and Anticancer Activity

Ethanol extract of the aerial parts of *S. larvata* was evaluated for its anti-inflammatory and antinociceptive activity. The data suggested that the extract has a significant antinociceptive effect [25].

The anticancer potential of *S. brasiliensis* leaf extracts were tested for cytotoxicity against four tumor cell lines: B16 (murine skin), HL-60 (human leukemia), MCF-7 (human breast) and HCT-8 (human colon). However, there was no inhibition of proliferative activity in the cell lines tested [35].

3.4. Antihyperlipidemic and Antihyperglycemic Effects

The hydroethanol extract of the roots of *S. brasiliensis* was evaluated for antihyperlipidemic and antihyperglycemic effect in mice fed a high-carbohydrate diet. It was observed that the extract represents a potential therapeutic resource in the treatment and prevention of dyslipidemias. There was a decrease in plasma glucose, cholesterol and triglyceride concentrations [36].

3.5. Allelopathic and Larvicidal Activity

Petroleum ether and methanol extracts, besides the fatty acids and methyl esters obtained from leaves of *S. brasiliensis*, were evaluated for the growth of radicles and hypocotyls of *Lactuca sativa* (lettuce) and *Allium cepa* (onion). Extracts, fatty acids and methyl esters inhibited the growth of radicals and hypocotyls of *A. cepa*. For *L. sativa*, leaf samples of *S. brasiliensis* presented diverse allelopathic activity with respect to radicle and hypocotyl growth. The tested substances gave significant allelopathic effects, showing promising results [37].

Another study of the allelopathic effect of ethanol extract and leaf fractions of *S. brasiliensis* was found. The extract and the fractions inhibited the growth of the hypocotyls and radicles of *A. cepa*. In the study with *L. sativa*, the samples presented stimulatory and inhibitory activity of the growth of the hypocotyls and radicles [28].

Petroleum ether extract, methanol extract, fatty acids and methyl esters of leaves of *S. brasiliensis* were evaluated for their cytotoxicity through a test with *Artemia salina*. The results showed that these substances are non-toxic since LC_{50} (lethal concentration) values ranged from 681.59 μg/mL to above 1000 μg/mL [29]. The study by Amado et al. [27], made with the leaves of *S. brasiliensis*, showed that the methanol extract and dichloromethane, ethyl acetate and hydromethanol fractions also do not present toxicity, since they presented a lethal concentration above 1000 μg/mL. A third study with leaves of *S. brasiliensis* presented results with an LC_{50} above 1000 μg/mL in the *A. salina* test [35].

The ethanol extract and the hexane, chloroform, ethyl acetate and hydroethanol fractions of the aerial parts of *S. larvata* were also not toxic to *A. salina* larvae [18].

Extracts and fractions of *S. brasiliensis* leaves were also submitted to a toxicity test on *Culex quinquefasciatus* larvae. The results showed LC_{50} values of 469.78 μg/mL to greater than 1000 μg/mL, indicating low toxicity on the larvae tested [27].

The ethanol extract of the aerial parts of *S. larvata* presented activity against larvae of *Aedes aegypti*, with LC_{50} = 225 μg/mL, a value that indicates a toxic effect on the larvae [18].

Conclusion

Based on the information provided, it is concluded that *Smilax* species need to be better investigated, due to the presence of compounds that are the basis for their already determined pharmacological activity. This review brought together research carried out by several scientists in order to provide a basis for future studies.

Acknowledgments

The authors are grateful to the Federal University of São João Del Rei (UFSJ), the National Council for Scientific and Technological Development (CNPq), Coordination of Improvement of Higher Level Personnel (CAPES) and the Foundation for Research Support of the State of Minas Gerais (FAPEMIG) for their fellowships.

References

[1] Marques, M. B. (2000). Pharmaceutical patents and the accessibility of drugs in Brazil. *R. Hist. Cienc. Saúde*, 7, 7-21.

[2] Simões, C. M. O., Schenkel, E. P., Gosmann, G., Mello, J. C. P., Mentz, L. A., Petrovick, P. R. (2010). *Farmacognosia, da planta ao medicamento*. [*Pharmacognosy, from plant to medicine*]. Porto Alegre: UFRGS.

[3] Guarim Neto, G., Morais, G. (2003). Medicinal plants resources in the Cerrado of Mato Grosso State, Brazil: a review. *Acta Bot. Bras.*, 17, 561-84.

[4] *The Plant List.* Version 1.1. (2013). Available from: http://www.theplantlist.org. Accessed on August 16, 2018.

[5] Andreata, R. H. P. (2006). *Smilacaceae* in the Biological Reserve of Poço das Antas, Rio de Janeiro, Brazil. *Rodriguésia*, 57, 647-657.

[6] Judd, W. S., Campbell, C. S., Kellogg, E. A., Stevens, P. F., Donoghue, M. J. (2009). *Sistemática vegetal: um enfoque filogenético.* [*Plant systematics: a phylogenetic approach*]. Porto Alegre: Artmed.

[7] Andreata, R. H. P. (2000). *Flora fanerogâmica da Ilha do Cardoso (São Paulo, Brasil).* [*Phanerogamic flora of Cardoso Island (São Paulo, Brazil)*]. São Paulo: Instituto de Botânica.

[8] Changwei, A., Tatsunori, H., Khanh, T. D., Upadhyay, A., Tawata, S. (2011). Antioxidant phenolic compounds from *Smilax sebeana* Miq. *Food Sci. Technol.*, 44, 1681-1686.

[9] Medeiros, M. F., Valle, L. S., Andreata, R. H. P. (2007). Histórico e o uso de Salsa parrilha (*Smilax* spp.) pelos boticários no Mosteiro de São Bento. [History and the use of "sarsaparilla" (*Smilax* spp.) by apothecaries in the Monastery of São Bento]. *Rev. Bras. Biocienc.*, 5, 27-29.

[10] Andreata, R. H. P. (2015). *Smilacaceae, Lista de Espécies da Flora do Brasil. Jardim Botânico do Rio de Janeiro.* [*Smilacaceae, List of Species of Flora of Brazil. Botanical Garden of Rio de Janeiro*]. Available from: http://floradobrasil.jbrj.gov.br/jabot/floradobrasil/FB32808. Accessed on August 16, 2018.

[11] Silva Júnior, A. A. (2006). *Essentia herba: Plantas bioativas.* [*Essentia herba: Bioactive plants*]. Florianópolis: Editora Epagri.

[12] Andreata, R. H. P. (1997). Revisão das espécies brasileiras do gênero *Smilax* Linnaeus (*Smilacaceae*). [Review of the Brazilian species of the genus *Smilax* Linnaeus (*Smilacaceae*)]. *Pesquisas-Botânica*, 47, 7-244.

[13] Martins, A., Bombo, B., Soares, A., Appezzato-da-Glória. (2013). Aerial stem and leaf morphoanatomy of some species of *Smilax*. *Braz. J. Pharmacogn.*, 23, 576-584.

[14] Alves, E. O., Mota, J. H., Soares, T. S., Vieira, M. C., Silva, C. B. D. A. (2008). Etnobotanical survey and medicinal plants characterization in forest fragments in Dourados-MS. *Cienc. Agrotec.*, 32, 651-658.

[15] Jiang, J., Xu, Q. (2003). Immunomodulatory activity of the aqueous extract from rhizome of *Smilax glabra* in the later phase of adjuvant-induced arthritis in rats. *J. Ethnopharmacol.*, 85, 53-59.

[16] Rugna, A., Polo, J., Evelson, P., Gurni, A. A., Llesuy, S., Wagner, M. L. (2003). Antioxidant activity in rhizomes from *Smilax campestris* Griseb. *Smilacaceae*. *Mol. Med. Chem.*, 1, 21-25.

[17] Soares, K., Martin, A., Montanari Júnior, I., Figueira, G. M., Zuchhi, M. I., Bajay, M. M., Appezzato-da-Glória, B. (2014). The sarsaparilla market in the state of São Paulo (Brazil) and the challenges of cultivation. *Braz. J. Pharmacogn.*, 24, 73-79.

[18] Hirota, B. C. K., Merino, F. J. Z., Verdam, M. C. S., Silva, C. B., Murakami, F. S., Miguel, O. G., Miguel, M. D. (2015). Larvicide and antifungal activities of sarsaparilla (*Smilax larvata*) extracts. *Int. J. Pharm. Pharm. Sci.*, 7, 308-311.

[19] Monteiro, M. H. D. A., Andreata, R. H. P. (1997). A pharmacological and botanical study of *Smilax quinquenervia* Vellozo from Brazil. Uso Y manejo de recursos Vegetales. In: *Memorias del Segundo Simposio Ecuatoriano de Etnobotánica Económica*. [Use and management of plant resources. In: *Memories of the Second Ecuadorian Economic Ethnobotany Symposium*].

[20] Demo, A., Petrankis, C., Kefalas, P., Boskou, D. (1998). Nutrient antioxidants in some herbs and Mediterranean plant leaves. *Food Res. Int.*, 31, 351-354.

[21] Almeida, S. P., Proença, C. E. B., Sano, S. M., Ribeiro, J. F. (1998). *Cerrado: espécies vegetais úteis*. [*Cerrado: useful plant species*]. Planaltina: Embrapa - CPAC.

[22] Alice, C. B., Vargas, V. M. F., Silva, G. A. A. B., Siqueira, N. C. S., Schapoval, E. E. S., Gleye, J., Henriques, J. A. P., Henriques, A. T. (1991). Screening of plants used in South Brazil folk medicine. *J. Ethnopharmacol.*, 35, 165-171.

[23] Soares, A. N. (2010). *Morfoanatomia, perfil químico e propagação de Smilax fluminensis Steud. (Smilacaceae).* [*Morphoanatomy, chemical profile and propagation of Smilax fluminensis Steud. (Smilacaceae)*]. Dissertation (Master of Science), Piracicaba.

[24] Breitbach, U. B., Niehues, M., Lopes, N. P., Faria, J. E. Q., Brandão, M. G. L. (2013). Amazonian Brazilian medicinal plants described by C.F.P. von Martius in the 19th century. *J. Ethnopharmacol.*, 147, 180-189.

[25] Hirota, B. C. K., Paula, C. S., Oliveira, V. B., Cunha, J. M., Schreiber, A. K., Campos, F. M. M., Barison, A., Miguel, O. G., Miguel, M. D. (2016). Phytochemical and antinociceptive, anti-inflammatory, and antioxidant studies of *Smilax larvata* (*Smilacaceae*). *Evid. Based Complementary Altern. Med.*, 1, 1-12.

[26] Martins, A. R. (2009). *Morfoanatomia dos órgãos vegetativos e perfil químico de espécies do gênero Smilax L. (Smilacaceae).* [*Morphoanatomy of vegetative organs and chemical profile of species of the genus Smilax L. (Smilacaceae)*]. Thesis (PhD in Plant Biology), Campinas.

[27] Amado, P. A., Castro, A. H. F., Alves, S. N., Silva, D. B., Carollo, C. A., Lima, L. A. R. S. (2019). Phenolic compounds: antioxidant and larvicidal potential of *Smilax brasiliensis* Sprengel leaves. *Nat. Prod. Res.*, doi: 10.1080/14786419.2018.1543678.

[28] Fonseca, J. C., Barbosa, M. A., Silva, I. C. A., Duarte-Almeida, J. M., Castro, A. H. F., Dos Santos Lima, L. A. R. (2017). Antioxidant and allelopathic activities of *Smilax brasiliensis* Sprengel (*Smilacaceae*). *S. Afr. J. Bot.*, 111, 336-340.

[29] Amado, P. A., Ferraz, V., Silva, D. B., Carollo, C. A., Castro, A. H. F., Lima, L. A. R. S. (2018). Chemical composition, antioxidant and cytotoxic activities of extracts from the leaves of *Smilax brasiliensis* Sprengel (*Smilacaceae*). *Nat. Prod. Res.*, 32, 610-615.

[30] Morais, M. I., Pinto, M. E. A., Araújo, S. G., Castro, A. H. F., Duarte-Almeida, J. M., Rosa, L. H., Rosa, C. A., Johann, S., Lima, L. A. R. S. (2014). Antioxidant and antifungal activities of *Smilax campestris* Griseb. (*Smilacaceae*). *Nat. Prod. Res.*, 28, 1275-1279.

[31] Battista, S. M., Garcia, G. F., Rugna, A. Z., Wagner, M. L., Gurni, A. A. (2007). Antifungal activity of extracts obtained from *Smilax campestris* Griseb. *Smilacaceae*. *Bol. Latinoam. Caribe Plant Med. Aromaticas*, 6, 330-331.

[32] Rodrigues, K., Ramos, D. F., Carrion, L. L., Cursino, L. M. C., Jefreys, M. F., Pedroza, L. S., Osório, M. I. C., Oliveira, J. L., Andrade, J. I. A., Fernandes, C. C., Nunez, C. V., Silva, P. E. A. (2014). Antifungal activity of Brazilian amazon plants extracts against some species of *Candida* spp. *Int. J. Pharm.*, 5, 445-453.

[33] Silva, L. I., Karuppusamy, A., Miyajima, F., Violante, I. M. P., Bieski, I. G. C., Balogun, S. O., Martins, D. T. O. (2017). Antimicrobial and antioxidant activities of selected plants used by populations from Juruena Valley, legal Amazon Brazil. *Int. Pharm. Pharm. Sci.*, 9, 179-191.

[34] Suffredini, I. B., Paciencia, M. L. B., Varella, A. D., Younes, R. N. (2006). Antibacterial activity of Brazilian Amazon plant extracts. *Braz. J. Infect. Dis.*, 10, 400-402.

[35] Santos Júnior, H. M., Oliveira, D. F., Carvalho, D. A., Pinto, J. M. A., Campos, V. A. C., Mourão, A. R. B., Pessoa, C., Moraes, M. O., Costa-Lotufo, L. V. (2010). Evaluation of native and exotic Brazilian plants for anticancer activity. *J. Nat. Med.*, 64, 231-238.

[36] Pereira, F. L., Oliveira, V. B., Viana, C. T. R., Campos, P. P., Silva, M. A. N., Brandão, M. G. L. (2015). Antihyperlipidemic and antihyperglycemic effects of the Brazilian salsaparrilhas *Smilax brasiliensis* Spreng. (*Smilacaceae*) and *Herreria salsaparrilha* Mart. (Agavaceae) in mice treated with a high-refined-carbohydrate containing diet. *Food Res. Int.*, 76, 366-372.

[37] Amado, P. A., Castro, A. H. F., Stein, V. C., Alves, S. N., Lima, L. A. R. S. (2019). Allelopathic, cytotoxic, genotoxic and larvicidal effects of *Smilax brasiliensis* Sprengel leaves. In: *Phytochemicals: Plant Sources and Potential Health Benefits.* New York: Nova Science Publishers.

In: Phytochemicals
Editor: Iman Ryan
ISBN: 978-1-53615-478-8

Chapter 6

FIRST PHYTOCHEMICAL STUDIES AND BIOLOGICAL ACTIVITIES OF *SMILAX FLUMINENSIS* STEUD

Juliana Costa Fonseca, Izabela Caputo Assis Silva, Jéssica Elvira Siqueira Medeiros, Ana Hortência Fonsêca Castro and Luciana Alves Rodrigues dos Santos Lima
Campus Centro-Oeste Dona Lindu, Universidade Federal de São João Del Rei, Divinópolis, Minas Gerais, Brazil

ABSTRACT

Smilax fluminensis is popularly known as japecanga or sarsaparilla. This species is a dioecious liana with a great capacity for shoot formation in the soil and primary branch formation in the subsoil. It has a wide geographic distribution, occurring over a range of environments from closed areas such as the Amazon Forest to open places such as the Pantanal. The adaptive success of the species can be attributed to its genotypic plasticity. Although the genus *Smilax* is not well studied, the use of *S. fluminensis* has been reported in folk medicine since antiquity, with antibacterial, antifungal, anti-inflammatory and hematoprotective activities, as well as in the treatment of syphilis and rheumatism. Here we aim to review the first phytochemical studies on this plant. Phytochemical work using the leaves of *S. fluminensis* indicated the presence of flavonoids, isolating and identifying quercetin-3-*O*-α-rhamnopyranoside (1-6)-*O*-β-D-glucopyranoside and quercetin-3-*O*-β-D-galactopyranoside. In another study, the presence of the following compounds was detected: coumarins, tannins, flavonoids, steroids, steroidal saponins glycosides, cardiotonic glycosides, and alkaloids. Chlorogenic acid, caffeic acid, rutin, *p*-coumaric acid, and *trans*-cinnamic acid were identified by high performance liquid chromatography (HPLC). The first study showing the facility for propagation of this species, through the use of seeds and remains of seedlings and shoots, corroborates its potential economic use in a sustainable way, constituting an alternative to reducing the predatory extraction of this native species, if its pharmacological potential is confirmed.

Keywords: *Smilacaceae*, *Smilax*, chemical composition, biological activities

1. The *Smilax fluminensis* Species

The *Smilacaceae* family is grouped into Monocotyledoneae in the order Liliales and comprises the five genera *Anikenton, Coprosmanthus, Heterosmilax, Nemexia,* and *Smilax* [1]. The *Smilacaceae* family includes approximately 310 species distributed in tropical and subtropical climates [2, 3].

In Brazil, only the genus *Smilax* is found, with approximately 33 species, of which 14 are unique to Brazil [4, 5]. The *Smilax* species are popularly known as sarsaparilla or japecanga, found at less humid forest edges, particularly in seasonal forests and forests [6]. *Smilax fluminensis* Steud (Figure 1) occurs in the states of Roraima, Pará, Bahia, Minas Gerais, Rio de Janeiro, São Paulo, and the Midwest, extending to Bolivia, Paraguay, and Argentina [7].

Figure 1. The species *Smilax fluminensis* Steud. A) herbal plant, B) leaf.

The roots and rhizomes of several *Smilax* species are used in traditional medicine for their antibacterial, antifungal, anti-inflammatory, and hepatoprotective effects, as well as for the treatment of syphilis [8-13].

Despite the pharmacological activities attributed to *Smilax* species, it is difficult to identify such plants taxonomically, due to similarities in their morphology, chemical composition, and popular names. The *Smilax* species are dioecious and take the form of vines, herbaceous vines, or rarely, subshrubs or shrubs. The leaves are simple and alternate, containing petioles with tendrils and primary acrodrome venation. The thickened stems are rhizophores, and the aerial stems are generally aculeate [5].

Therefore, to aid in distinguishing *S. fluminensis* from the other species of the genus, it is necessary to observe anatomical characteristics, such as the ultrastructure of the leaf cuticle [14].

In *S. fluminensis*, the leaf blades are long and have a coriaceous consistency, with the shape of ropes, presenting a basal nerve with five large, central veins. The walls of the adaxial and abaxial epidermis are sinuous. The leaf is hypostomatic, and there are few stomata on the adaxial surface, which are anisocytic and paracytic. The leaf margin of this species presents a non-stratified epidermis with smaller and more rounded cells than the rest of the leaf blade and thick cuticles. The layers of cells beneath the epidermis have a narrow lumen and thick, lignified walls. Other cells in the margin may have phenolic compounds, with idioblasts containing raphides. The lateral vascular bundles are always surrounded by lignified sclereid cells and fibers [10].

The stems have several spiny structures in the internodal region and do not have vasculature in the internodal and nodal regions. In the third internodal region of the aerial rod, *S. fluminensis* has thin-walled cells and larger lumens in the starch sheath [10].

The flowers of *S. fluminensis* are grouped into inflorescences of the umbeliform type, with a greenish, trimeric perigony. The tepalas of the two whorls are arranged alternately; the three inner parts are oblong, and the other parts are elliptical. The staminate flowers present six or, rarely, seven free stamens, with linear anthers and longitudinal dehiscence. In the pistil flowers, there are six filiform staminodes, papillose at the apex and the margins. The ovary is superimposed, tricarpelar, and trilocular, with a seminal rudiment (ovule) per locule. The carpels are stigmatized with the papillary adaxial face [15].

The regions corresponding to the cortex and medulla are composed of parenchyma formed of isodiametric cells. Histochemical tests showed the presence of phenolic substances in the epidermal cells; lipids and starch in the parenchyma cells of the cortex and the marrow; and idioblasts with calcium oxalate and mucilage raphides [15].

2. Phytochemical Studies of *Smilax fluminensis*

There are few studies on this species, especially those related to the phytochemistry of *S. fluminensis*. Lucas et al. [16] detected in the leaves of *S. fluminensis* the presence of phenolic compounds, coumarins, tannins, flavonoids, and steroids in the ethanol extract; steroidal saponins glycosides, cardiotonic glycosides, and alkaloids in the dry extract; and phenolic compounds and coumarins in the aqueous extract.

The chemical profile of *S. fluminensis* analyzed with high performance liquid chromatography (HPLC) revealed the presence of chlorogenic acid, caffeic acid, rutin, *p*-coumaric acid, and *trans*-cinnamic acid in the extracts of roots, rhizophores, and leaves [17]. Martins [18] verified that chlorogenic, *p*-coumaric, and ferulic acids are common in the secondary metabolism of *S. brasiliensis*, *S. campestris*, *S. cissoides*, *S. fluminensis,* and *S. polyantha* species.

Petrica et al. [19] isolated and elucidated two chemical constituents present in *S. fluminensis.* The isolated compounds were analyzed using HPLC and ^{1}H and ^{13}C NMR (nuclear magnetic resonance) and identified as quercetin-3-*O*-α-L-rhamnopyranoside (1-6)-*O*-β-D-glucopyranoside and quercetin-3-*O*-β-D-galactopyranoside [19].

3. BIOLOGICAL ACTIVITIES OF *SMILAX FLUMINENSIS*

No studies on the biological activities of *S. fluminensis* were found. However, Lucas et al. [16] detected in the leaves of *S. fluminensis* several compounds, such as coumarins, tannins, flavonoids, steroids, saponins, and alkaloids, which may be correlated with the diuretic and antioxidant effects found in other *Smilax* species.

In the same study, the foam index made it possible to quantify saponins in leaves at 153, indicating that this amount of saponins does not present a danger to human health. Tests of total ash and acid-insoluble ash expressed as 3.23% of total ash and 2.66% of acid-insoluble ash also suggest that the plant has low toxicity [16].

The roots and rhizomes of several plants of the genus *Smilax* are used in traditional medicine due to their biological activities [8-13]. However, collecting these parts of the plant may lead to the extinction of the species. Thus, the propagation of medicinal species is of great importance because increasing cultivation and discouraging extraction promotes the sustainable management of these species [20]. A study aimed at analyzing the propagation capacity of *S. fluminensis* found significant data on this species [21].

Seed germination of *S. fluminensis* was monitored weekly for 30 weeks. Germination was recorded after the third week of the experiment based on aerial shoot germination. For all combinations of temperature and light studied, no significant difference in germination was found between light and dark, i.e., light did not interfere with germination (neutral photoblastic).

The percent germination recorded at the end of the experiment was 73% and 85% for treatment at 30°C in light and darkness, respectively; 80% and 66% for treatment between 20 and 30°C; 24% and 33% for treatment between 20 and 35°C; and 65% at 25°C [17, 21].

In a second experiment, in the aerial branches and underground branches of plagiotrophic distribution removed to make the stakes, a swollen nodal region was always left. Plant sprouting started 30 days after the experiment, evidenced by the development of new shoots. At the end of the experiment, all plants had new shoots of different sizes and stages of development. The largest branches were about 1.80m long. Each plant had on average 4.05 new shoots. Shoot development was evident from the fifth month after the initiation of the experiment, that is, at the favorable time for its development in the Brazilian Cerrado, because of the increase in temperature and precipitation [17, 21].

Thus, the ability to propagate the species through the use of seeds and remains of seedlings and shoots after the removal of aerial stems in *S. fluminensis* corroborates their potential for sustainable economic use, constituting an alternative that will reduce the predatory extraction of native species [21].

CONCLUSION

Initial studies with the species *S. brasiliensis* have shown the presence of several classes of compounds, such as phenolic compounds, coumarins, tannins, flavonoids, steroids, steroidal saponins glycosides, cardiotonic glycosides, and alkaloids, which can be correlated with biological activities already described for other species of *Smilax*. The high rate of propagation of the species suggests their potential for sustainable economic use, if its pharmacological potential is confirmed.

ACKNOWLEDGMENTS

The authors are grateful to the Federal University of São João Del Rei (UFSJ), National Council for Scientific and Technological Development (CNPq), Coordination of Improvement of Higher Level Personnel (CAPES) and Foundation for Research Support of the State of Minas Gerais (FAPEMIG) for their fellowships.

REFERENCES

[1] *The Plant List.* Version 1.1. (2013). Available from: http://www.theplantlist.org. Accessed on August 16, 2018.

[2] Andreata, R. H. P. (2006). *Smilacaceae* in the Biological Reserve of Poço das Antas, Rio de Janeiro, Brazil. *Rodriguésia*, 57, 647-657.

[3] Judd, W. S., Campbell, C. S., Kellogg, E. A., Stevens, P. F., Donoghue, M. J. (2009). *Sistemática vegetal: um enfoque filogenético*. [*Plant systematics: a phylogenetic approach*]. Porto Alegre: Artmed.

[4] Andreata, R. H. P. (2015). *Smilacaceae. Lista de Espécies da Flora do Brasil.* [*Smilacaceae. List of Species of Flora of Brazil*]. Jardim Botânico do Rio de Janeiro. Available from: http://floradobrasil.jbrj.gov.br/jabot/floradobrasil/FB32808. Accessed on August 16, 2018.

[5] Andreata, R. H. P. (1997). Revisão das espécies brasileiras do gênero *Smilax* Linnaeus (*Smilacaceae*). [Review of the Brazilian species of the genus *Smilax* Linnaeus (*Smilacaceae*)]. *Pesquisas-Botânica*, 47, 7-244.

[6] Souza, V. C., Lorenzi, H. (2008). *Botânica Sistemática: guia ilustrado para identificação das famílias de angiospermas da flora brasileira, baseado em APGII.* [*Systematic Botany: illustrated guide for identification of angiosperm families of the Brazilian flora, based on APGII*]. Nova Odessa: Instituto Plantarum.

[7] Andreata, R. H. P. (2015). Flora of the Serra do Cipó, Minas Gerais: *Smilacaceae*. [Flora of the Serra do Cipó, Minas Gerais: *Smilacaceae*]. *Bol. Bot. Univ. São Paulo*, 33, 39-44.

[8] Xu, J., Li, X., Zhang, P., Li, Z. L., Wang, Y. (2005). Antiinflammatory constituents from the roots of *Smilax bockii* Warb. *Arch. Pharm. Res.*, 28, 395-399.

[9] Mandal, S. C., Jana, G. K, Das, S., Sahu, R., Venkidesh, R., Dewanjee, S. (2008). Hepatoprotective and antioxidant activities of *Smilax chinensis* L. root. *Pharmacologyonline*, 2, 529-535.

[10] Martins, A. R., Bombo, A. B., Soares, A. N., Appezzato-da-Glória, B. (2013). Aerial stem and leaf morphoanatomy of some species of *Smilax*. *Braz. J. Pharmacogn.*, 23, 576-584.

[11] Alves, E. O., Mota, J. H., Soares, T. S., Vieira, M. C., Silva, C. B. (2008). Etnobotanical survey and medicinal plants characterization in forest fragments in Dourados-MS. *Cienc. Agrotec.*, 32, 651-658.

[12] Rugna, A., Polo, J., Evelson, P., Gurni, A. A., Llesuy, S., Wagner, M. L. (2003). Antioxidant activity in rhizomes from *Smilax campestris* Griseb. *Smilacaceae*. *Mol. Med. Chem.*, 1, 21-25.

[13] Soares, M. K. M., Martins, A. R., Montanari Junior, I., Figueira, G. M., Zuchhi, M. I., Bajay, M. M., Appezzato-da-Glória, B. (2014). The sarsaparilla market in the state of São Paulo (Brazil) and the challenges of cultivation. *Braz. J. Pharmacogn.*, 24, 73-79.

[14] Ferrufino-Acosta, L. (2014). Taxonomic revision of the genus *Smilax* (*Smilacaceae*) in Central America and the Caribbean Islands. *Willdenowia*, 40, 227-280.

[15] Souza, R. C. O. S., Toni, K. L. G., Andreata, R. H. P., Costa, C. G. (2005). Floral anatomy and vascular tissue in *Smilax fluminensis* Steudel (*Smilacaceae*). *Rodriguésia*, 56, 107-121.

[16] Lucas, J., Morinigo, P., Almeida, J. M., Coelho, R. M., Scheleder, E. J. D. (2010). Análise fitoquímica, índice de espuma e índice de cinzas totais e insolúveis em ácido das folhas de *Smilax fluminensis*. [Phytochemical analysis, foam index and total ash content and insoluble in acid of the *Smilax fluminensis* leaves]. *Anais do 1º Seminário Interno de Iniciação Científica*. Available from: http:// repositorio.pgsskroton.com.br/handle/123456789/377. Accessed on October 16, 2018.

[17] Soares, A. N. (2010). *Morfoanatomia, perfil químico e propagação de Smilax fluminensis Steud. (Smilacaceae).* [*Morphoanatomy, chemical profile and propagation of Smilax fluminensis Steud. (Smilacaceae)*]. Dissertation (Master of Science), Piracicaba.

[18] Martins, A. R. (2009). *Morfoanatomia dos órgãos vegetativos e perfil químico de espécies do gênero Smilax L. (Smilacaceae).* [*Morphology of vegetative organs and chemical profile of species of the genus Smilax L. (Smilacaceae)*]. Thesis (PhD in Biological Sciences), Piracicaba.

[19] Petrica, E. E. A., Sinhorin, A. P., Sinhorin, V. D. G., Júnior, G. M. V. (2014). First phytochemical studies of japecanga (*Smilax fluminensis*) leaves: flavonoids analysis. *Braz. J. Pharmacogn.*, 24, 443-445.

[20] Maroni, B. C., Stasi, L. C., Machado, S. R. (2006). *Plantas medicinais do Cerrado de Botucatu.* [*Medicinal plants of Cerrado of Botucatu*]. São Paulo: Fundação Editora Unesp.

[21] Soares, A. N., Novembre, A. D. L. C., Martins, A. R., Piedade, S. M. S., Appezzato-da-Glória, B. (2011). Propagation studies in *Smilax fluminensis* Steud. (*Smilacaceae*). *Cienc. Rural*, 41, 1762-1768.

In: Phytochemicals
Editor: Iman Ryan

ISBN: 978-1-53615-478-8

Chapter 7

PHYTOCHEMICAL ASPECTS AND BIOLOGICAL PROPERTIES OF *SMILAX BRASILIENSIS* SPRENGEL

Izabela Caputo Assis Silva, Juliana Costa Fonseca, Jéssica Elvira Siqueira Medeiros, Paula Avelar Amado, Ana Hortência Fônseca Castro and Luciana Alves Rodrigues dos Santos Lima

Campus Centro-Oeste Dona Lindu, Universidade Federal de São João Del Rei, Divinópolis, Minas Gerais, Brazil

ABSTRACT

Smilax brasiliensis Sprengel is a monocotyledon belonging to the family *Smilacaceae*. It is a native species of the Brazilian Cerrado and presents itself as a shrub, a climbing vine, or even a tree, being popularly known as japecanga or sarsaparilla. In folk medicine, it is used for the treatment of sexually transmitted diseases, cutaneous conditions, and inflammatory diseases and as a diuretic, diaphoretic, and depurative agent, the roots, and rhizomes being the main parts used. This review will report on the phytochemical aspects and biological activities of the species *S. brasiliensis*. Phytochemical investigations have already been carried out with different parts of the plant, such as the leaves, roots, rhizomes, stems, and branches. Studies with extracts and fractions of this species have demonstrated the presence of phenolic compounds, such as rutin, quercetin, and chlorogenic acid, and apolar compounds, especially fatty acids, with a high proportion of linolenic, linoleic, and palmitic acids. Other classes of secondary metabolites responsible for the diversity of biological actions in the plant have also been identified, among them coumarins, alkaloids, flavonoids, condensed tannins, steroids, triterpenoids, essential oils, and saponins, the latter being the main class of secondary metabolites found in the *Smilax* genus. Saponins are compounds with a steroidal structure and are of great industrial interest, since they have applicability in the production of anti-inflammatory drugs and contraceptives. As *S. brasiliensis* is still a little-studied species, only a few reports of its biological activities were found, such as antibacterial, antifungal, antioxidant, allelopathic, antihyperglycemic, antihyperlipidemic, and anticancer effects. No toxic effects on *Artemia salina* were observed, and it was not genotoxic. Considering that this species is not well explored with regard to its biological actions and the chemical compounds present in the same, *S. brasiliensis* can be an alternative source of pharmacological activities for application in the improvement of health and the environment.

Keywords: chemical constitution, *Smilax*, japecanga, medicinal plant, cerrado

1. THE *SMILAX* GENUS

The *Smilacaceae* family comprises about 310 species found in regions of tropical, subtropical and, to a lesser extent, temperate climates [1]. In Brazil, there are 33 described species, 19 of them endemic [2]. Plants of the genus *Smilax* are monocotyledonous and dioecious, with the appearance of a creeper [3]. They are known as japecanga or sarsaparilla [4], the latter name being attributed to their depurative effect [5].

In popular medicine, plants of this genus are used in the treatment of sexually transmitted diseases (STDs), skin diseases (abscesses, boils, psoriasis), rheumatism, swelling, toothache, dysentery [6], rheumatoid arthritis, low back pain, gout, tumors, and inflammatory diseases; as diuretic and detoxifying agents [7]; and for eye pain [8].

Several biological actions are reported for the *Smilax* genus, among them antioxidant, allelopathic, larvicidal [9, 10, 11], antimicrobial, analgesic, antidiabetic, cytotoxic, cytoprotective, hepatoprotective, anti-inflammatory, pesticidal [6], antitumoral [8], and immunomodulatory [12] activities.

This variety of effects is due to the great diversity of chemical constituents present in plants of the genus *Smilax*. Among these constituents, there is a major presence of saponins and a great interest in isolating them due to their industrial and pharmaceutical applications [13].

2. The Species *Smilax brasiliensis*

Smilax brasiliensis Sprengel (Figure 1) is a monocotyledon of the *Smilacaceae* family [14], native to the Brazilian Cerrado [15]. It is characterized by oval leaves with three central veins [16], stems with spines, a continuous sclerenquimic ring throughout the length of the stalk, the presence of starch grains around the vascular fecture between internodes [4], greenish flowers, green to wine and black globose berries, and reddish seeds [3].

Figure 1. The species *Smilax brasiliensis* Sprengel. A) herbal plant with leaves and vine aspect; B) inflorescences.

The species has not yet been well studied, and few reports exist on its biological properties and chemical constitution. Among its actions, the following stand out: antioxidant, allelopathic [9, 10], antimicrobial [17, 18], anticancer [19], antihyperglycaemic, and antihyperlipidemic [5, 20].

3. The Chemical Constitution of *Smilax brasiliensis*

Few studies have reported the chemical profile of *S. brasiliensis* Sprengel. In a study using high-performance liquid chromatography coupled with a diode array detector (HPLC-UV-PDA) to identify the chemical composition of *Smilax* species, including *S. brasiliensis,* it was possible to identify chlorogenic acid, caffeic acid, rutin, *p*-coumaric acid, ferulic acid, and *trans*-cinnamic acid in methanol extracts of roots, rhizophores, and leaves [21].

Another study using a crude extract and fractions evaluated the profile of *S. brasiliensis* by thin layer chromatography (TLC), and it was possible to observe the presence of chlorogenic acid and saponins (such as parillin and sarsaparilloside) [5, 22]. In analyses carried out by high-performance liquid chromatography coupled to ultraviolet (HPLC-UV), substances of high polarity were detected, consistent with the profile found in TLC of phenolic substances such as chalcones and saponins [5].

Chlorogenic acid and rutin were also identified in an ethanol extract and fractions of leaves of *S. brasiliensis* by TLC [9]. The dichloromethane fraction was fractionated and analyzed by gas chromatography coupled to mass spectrometry (GC/MS), and 10 compounds were detected, among them methyl laurate, methyl-12-methyl tridecanoate, methyl palmitate, methyl oleate, and methyl cyclopentaneundecanoate [16] (Figure 2).

The fatty acid methyl esters obtained from ethereal and methanol extracts of leaves of this species were analyzed by gas chromatography (GC) and gas chromatography with a flame ionization detector (GC/FID), and a similar profile was detected for both extracts, consisting of methyl laurate, methyl myristate, methyl palmitate, methyl 9-hexadecenoate, methyl stearate, methyl oleate, methyl linoleate, methyl linolenate, methyl eicosanoate, and methyl docosonoate. Some constituents were detected only in ethereal extract as methyl margarate, methyl tetracosanoate, and methyl octacosanoate. A higher proportion of unsaturated esters was observed in both extracts [10, 23]. In the same study, a TLC analysis was performed for flavonoids, detecting detected rutin and chlorogenic acid [23] (Figure 2).

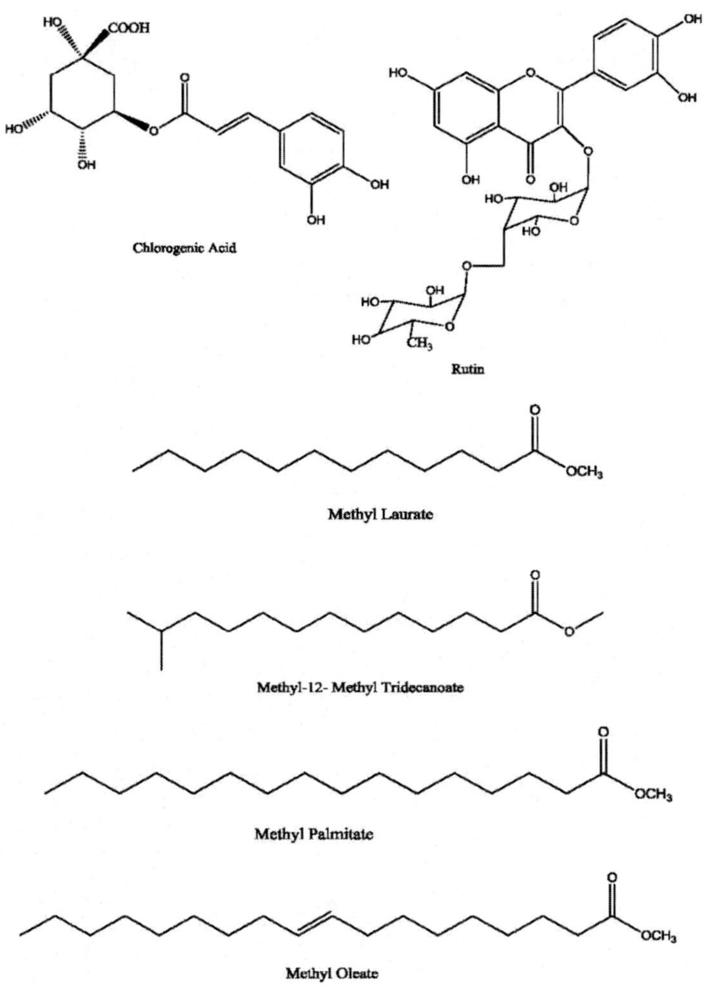

Figure 2. (Continued).

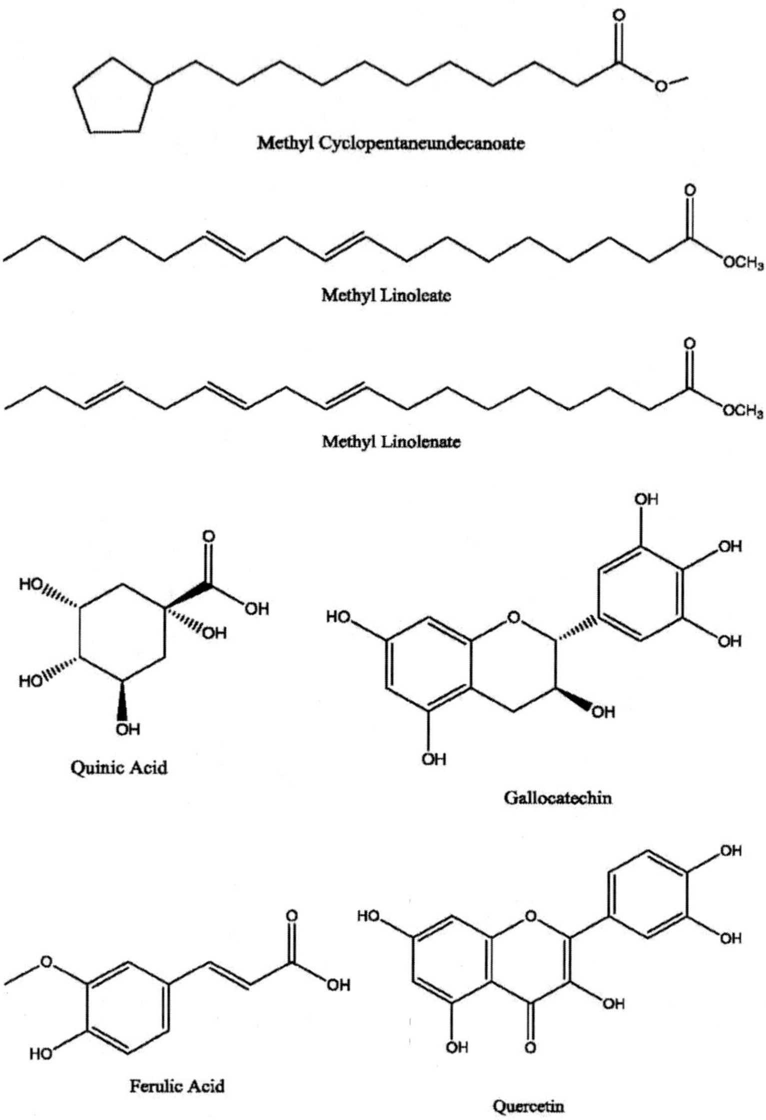

Figure 2. Structures of compounds identified in *S. brasiliensis.*

Other metabolites identified in the extract and fractions from *S. brasiliensis* [11] analyzed by liquid chromatography coupled to a diode array detector and mass spectrometer (LC-DAD-MS) included chlorogenic acid, flavonoids, and phenylpropanoids. The constituents identified were quinic acid, gallocatechin, 3-*O*-E-caffeoyl quinic acid, *O*-feruloyl quinic acid, *O*-caffeoyl shikimic acid, ferulic acid, 3-*O*-β-galactopyranosyl quercetin, 3-*O*-β-glicopyranosyl quercetin, quercetin, naringenin, and kaempferol (Figure 2).

4. Biological Activities of *Smilax brasiliensis*

4.1. Antimicrobial Activity

A study of 26 plants selected according to their medicinal use by the population of the Juruena Valley in Mato Grosso included among them *S. brasiliensis*. The antimicrobial potential of the hydroethanolic extract from branches and leaves was evaluated against gram-negative and gram-positive bacteria and fungi by minimum inhibitory concentration (MIC). The species presented moderate activity against *Enterococcus faecalis* and weak activity against *Aspergillus fumigatus*, with MIC values of 400 and 800 μg/mL, respectively [18].

Dichloromethane, methanol, and aqueous extracts were obtained from plants collected in regions of the Amazon Forest, among these *S. brasiliensis*. Extracts of roots, leaves, and branches were obtained and evaluated for antifungal action by determination of the minimum inhibitory concentration (MIC) against *Candida albicans, Candida glabrata,* and *Candida parapsilosis* strains. Only the dichloromethane extract from branches was inactive against the yeasts tested. The best activity was observed in cultures of *C. albicans* and *C. parapsilosis* treated with dichloromethane extract roots, with an MIC value of 0.06 μg/mL, demonstrating the highest inhibitory effect [17].

4.2. Anticancer Activity

S. brasiliensis was evaluated for anticancer activity, along with 50 species collected in Minas Gerais, and the methanol extract of this species was evaluated for its capacity to inhibit cell proliferation in four tumor cell lines: murine skin, human leukemia, human breast, and human colon. *S. brasiliensis* inhibited proliferation of human cell lines, and the highest effect was in human breast cells, showing 77.9% inhibition. In leukemia and human colon cells, proliferation was decreased by the action of the extract by 56.3% and 53.6%, respectively [19].

4.3. Antioxidant Activity

With the increase in diseases associated with the onset of oxidative stress, the search for natural substances that inhibits, neutralizes, or diminishes the action of free radicals has grown. Species with few studies, such as *S. brasiliensis*, have also been the focus of research.

Some reports were found on *S. brasiliensis*. Ethanol extract and fractions of leaves had an inhibitory effect on the 1,1-diphenyl-2-picryl-hydrazila (DPPH) free radical, with a high antioxidant potential, scavenging 50% of free radicals (IC_{50}) between 2.73 and 11.75 µg/mL, less than of the standard 2,6-di-tert-butyl-4-methylphenol (BHT) [9]. In another study, ethereal and methanol extracts and their fatty acids and methyl esters inhibited free radical DPPH, and IC_{50} values were 0.240-0.875 µg/mL [10, 23]. Amado et al. [11] evaluated the antioxidant effect of methanol extract and fractions of leaves, with small IC_{50} values (0.70-14.01 µg/mL).

In another study, 26 species, among these *S. brasiliensis*, were evaluated for antioxidant activity by DPPH, ferric reducing antioxidant power (FRAP), and nitric oxide (NO) assays. IC_{50} values were 171.83, >800, and >800 µg/mL, respectively. The results were correlated with the total phenolic compounds, flavonoids, and coumarins determined, and the contents were 1.47 ± 0.09, 0.32 ± 0.00, and 1.14 ± 0.00 mg/g, expressed as milligrams of rutin, tannic acid, and coumarin, respectively [18].

4.4. Allelopathic Activity

The allelopathic effect for the ethanol extract and fractions of *S. brasiliensis* on the seeds of *Allium cepa* (onion) and *Lactuca sativa* (lettuce) were evaluated. Growth inhibition of the radicles and hypocotyls of *A. cepa* at 125, 250, and 500 μg/mL was observed. For *L. sativa*, the effect was diversified, presenting inhibition and growth stimulus, for radicles and hypocotyls. The dichloromethane fraction of the leaves promoted greater inhibition of the growth of hypocotyls and radicles [9]. The allelopathic potential of ethereal and methanol extracts and their fatty acids and methyl esters of fatty acids (FAME) were also evaluated. The samples had an inhibitory effect on the growth of onion radicles and hypocotyls, at the three concentrations tested (250, 500, and 1000 μg/mL), and a diversified effect on lettuce seed growth [23, 25].

4.5. Cytotoxicity and Genotoxicity Effects

Chemical compounds with biological applicability may promote genotoxic and mutagenic effects [24]. Thus, the ethereal and methanol extracts of *S. brasiliensis* and its fatty acids and FAMEs were evaluated for cytogenetic effects on development of onion radicles at concentrations of 0, 250, 500, and 1000 μg/mL. The parameters evaluated were the mitotic index (MI), necrosis index (IN), and index of abnormalities (AI) [23, 25].

The samples of *S. brasiliensis* evaluated presented no cytotoxicity or genotoxic effects. Thus, it is possible to observe that the species *S. brasiliensis*, used in the treatment of many diseases, is safe since its extracts, fatty acids, and esters do not produce negative effects on human health [23, 25].

4.6. Toxicity and Larvicidal Effect

The 51 species tested for anticancer activity were also evaluated for lethality on *Artemia saline* Leach. *Artemia* is a brine shrimp used for toxicity determinations of substances according to its mortality. Methanol extract of *S. brasiliensis* presented a high lethal concentration to kill 50% (LC_{50} > 1000 μg/mL), demonstrating that the extract has a low toxic effect [19].

Another study evaluated the toxicity of the ethereal extract and their fatty acids and FAMEs obtained from leaves of *S. brasiliensis*. Fatty acids and FAMEs presented LC_{50} values below 1000 μg/mL on *A. salina*, however, suggesting low toxicity [10, 23].

Amado et al. [11] evaluated cytotoxic and larvicidal effect of methanol extract and fractions of leaves. The LC_{50} values on *C. quinquefasciatus* were 469.78 and 986.47 μg/mL for the dichloromethane fraction and methanol extract, respectively. The samples were not cytotoxic to *A. salina*, indicating selectivity against *C. quinquefasciatus* larvae [11].

4.7. Metabolic Activity

Sarsaparillas are several plants of different genera that act by cleaning the blood. In this study, BALB/c mice were submitted to a diet rich in refined carbohydrate (HC) from the 9th to the 12th weeks, receiving the HC diet supplemented by crude extract of *S. brasiliensis* in two dosages (100 and 200 mg/kg), separately. As a control (C), one group received only ration from the 1st to the 12th week. The parameters evaluated were food intake, body weight gain, adiposity index (IA), morphological analysis of adipocytes of adipose epididymal tissue, analysis of liver weight and morphology, and serological parameters (fasting glucose, total cholesterol, and triglycerides) [5, 20].

The animal feed intake of the groups treated with HC and HC diets, supplemented with crude *S. brasiliensis* extract at concentrations of 100 mg/kg (HCSI) and 200 mg/kg (HCSII) were higher than in the control group (C). Regarding body weight, animals with diet C, as well as diets HC and HC supplemented with extract, presented similar body weights. Animals treated with diet HC and HCSII (200 mg/kg) presented higher AI than group C and HCSI (100 mg/kg) [5, 20].

In the histological analysis of epididymal adipose tissue, animals from the HC and HCSI groups showed an increase in adipocyte area when compared with group C. The HCSII group presented an area of adipocytes similar to those in group C. All groups evaluated showed no hepatic increase or morphological changes [5, 20].

The serological analysis showed that the group treated with a HC diet showed an increase in the plasma concentrations of total cholesterol, glucose, and triglycerides. With respect to the concentrations of total cholesterol and glucose, it was possible to observe that the crude extract of *S. brasiliensis* in the highest dosage tested, 200 mg/kg, significantly decreased these plasma concentrations relative to HC. Triglyceride concentrations in all supplemented HC groups were reduced significantly relative to the HC group. The HCSII group presented a significant reduction relative to the control C. The data from this study demonstrate that species have a high hypolipidemic potential and that the effects can be attributed to the phenolic substances and saponins identified in the study [5, 20].

CONCLUSION

Based on the results, it is concluded that *S. brasiliensis* merits further investigation, due to the presence of compounds which may be responsible for the biological activities evaluated. This review brought together research carried out by several scientists in order to provide a basis for future studies of this species, *S. brasiliensis*.

ACKNOWLEDGMENTS

The authors are grateful to the Federal University of São João Del Rei (UFSJ), National Council for Scientific and Technological Development (CNPq), Coordination of Improvement of Higher Level Personnel

(CAPES) and Foundation for Research Support of the State of Minas Gerais (FAPEMIG) for their fellowships.

References

[1] Qi, Z., Cameron, K. M., Li, P., Zhao, Y., Chen, S., Chen, G., Fu, C. (2013). Phylogenetics, character evolution, and distribution patterns of the greenbriers, *Smilacaceae* (Liliales), a near-cosmopolitan family of monocots. *Bot. J. Linean Soc.*, 173, 535-548.

[2] Andreata, R. H. P., Watanabe, M. T. C. (2018). Flora of the *canga* of the Serra dos Carajás, Pará, Brazil: *Smilacaceae*. *Rodriguésia*, 69, 245-250.

[3] Andreata, R. H. P. (2003). *Smilacaceae*. In: Wanderley, M. G. L., Shepherd, G. J., Giulietti, A. M., Melhem, T. S. *Flora fanerogâmica do estado de São Paulo.* [*Phanerogamic Flora of the São Paulo State*]. São Paulo: FAPESP, RiMa, 3, 323-332.

[4] Martins, A. R., Abreu, A. G., Bajay, M. M., Villela, P. M. S., Batista, C. E. A., Monteiro, M., Alves-Pereira, A., Figueira, G. M., Pinheiro, J. B., Appezzato-da-Glória, B., Zucchi, M. I. (2013). Development and characterization of microsatellite markers for the medicinal plant *Smilax brasiliensis* (*Smilacaceae*) and related species. *Appl. Plant Sci.*, 1, 1-3.

[5] Pereira, F. L. (2013). *Potencial das raízes de Smilax brasiliensis Spreng. (Smilacaceae) e Herreria salsaparilha Mart. (Agavaceae) no tratamento de alterações metabólicas, induzidas por dieta em camundongos balb/c.* [*Potential of the roots of Smilax brasiliensis Spreng. (Smilacaceae) and Herreria salsaparilha Mart. (Agavaceae) in the treatment of metabolic changes induced by diet in balb/c mice*]. Dissertation (Food Science), Belo Horizonte.

[6] Dhanya Shree, V. S., Ayesha, A., Saema Noorain, G. K., Sahana, B. K., Prashith Kekuda, T. R. (2018). Preliminary phytochemical analysis, antimicrobial and antioxidant activity of *Smilax zeylanica* L. (*Smilacaceae*). *J. Drug Deliv. Ther.*, 8, 237-243.

[7] Zhang, Q. F., Cheung, H. Y., Zeng, L. B. (2013). Development of HPLC fingerprint for species differentiation and quality assessment of *Rhizoma Smilacis Glabrae*. *J. Nat. Med.*, 67, 207-211.

[8] Nho, K. J., Chun, J. M., Kim, H. K. (2015). Anti-metastatic effect of *Smilax china* L. extract on MDA-MB-231 cells. *Mol. Med. Rep.*, 11, 499-502.

[9] Fonseca, J. C., Barbosa, M. A., Silva, I. C. A., Duarte-Almeida, J. M., Castro, A. H. F., Dos Santos Lima, L. A. R. (2017). Antioxidant and allelopathic activities of *Smilax brasiliensis* Sprengel (*Smilacaceae*). *S. Afr. J. Bot.*, 111, 336-340.

[10] Amado, P. A., Ferraz, V., Silva, D. B., Carollo, C. A., Castro, A. H. F., Lima, L. A. R. S. (2018). Chemical composition, antioxidant and cytotoxic activities of extracts from the leaves of *Smilax brasiliensis* Sprengel (*Smilacaceae*). *Nat. Prod. Res.*, 32, 610-615.

[11] Amado, P. A., Castro, A. H. F., Alves, S. N., Silva, D. B., Carollo, C. A., Lima, L. A. R. S. (2019). Phenolic compounds: antioxidant and larvicidal potential of *Smilax brasiliensis* Sprengel leaves. *Nat. Prod. Res.*, doi: 10.1080/14786419.2018.1543678.

[12] Xu, W. A., Yin, L., Pan, H. Y., Shi, L., Xu, L., Zhang, X., Duan, J. A. (2013). Study on the correlation between constituents detected in serum from Rhizoma *Smilacis Glabrae* and the reduction of uric acid levels in hyperuricemia. *J. Ethnopharmacol.*, 150, 747-754.

[13] Rugna, A., Polo, J., Evelson, P., Gurni, A. A., Llesuy, S., Wagner, M. L. (2003). Antioxidant activity in rhizomes from *Smilax campestris* Griseb. *Smilacaceae*. *Mol. Med. Chem.*, 1, 21-25.

[14] Martins, A. R., Bombo, A. B., Soares, A. N., Appezzato-da-Glória, B. (2013). Aerial stem and leaf morphoanatomy of some species of *Smilax*. *Braz. J. Pharmacogn.*, 23, 576-584.

[15] Martins, A. R., Putz, N., Soares, A. N., Bombo, A. B., Appezzato-da-Glória, B. (2010). New approaches to underground systems in Brazilian *Smilax* species (*Smilacaceae*). *J. Torrey Bot. Soc.*, 137, 220-235.

[16] Barbosa, M. A. (2015). *Anatomia foliar e avaliação das atividades antioxidante e alelopática de folhas de Smilax brasiliensis Spreng.*

(Smilacaceae). [*Leaf anatomy and evaluation of the antioxidant and allelopathic activities of leaves of Smilax brasiliensis Spreng. (Smilacaceae)*]. Dissertation (Pharmaceutical Sciences), Divinópolis.

[17] Rodrigues, K., Ramos, D. F., Carrion, L. L., Cursino, L. M. C., Jefrey, M. F., Pedroza, L. S., Osorio, M. I. C., Oliveira, J. L., Andrade, J. I. A., Fernandes, C. C., Nunez, C. V., Silva, P. E. A. (2014). Antifungal activity of Brazilian Amazon plants extracts against some species of *Candida* spp. *Int. J. Phytopharmacol.*, 5, 445-453.

[18] Silva, L. I., Karuppusamy, A., Miyajima, F., Violante, I. M. P., Bieski, I. G. C., Balogun, S. O., Martins, D. T. O. (2017). Antimicrobial and antioxidant activities of selected plants used by populations from Juruena valley, legal Amazon, Brazil. *Int. J. Pharm. Pharm. Sci.*, 9, 179-191.

[19] Santos Júnior, H. M., Oliveira, D. F., Carvalho, D. A., Pinto, J. M. A., Campos, V. A. C., Mourão, A. R. B., Pessoa, C., Moraes, M. O., Costa-Lotufo, L. V. (2010). Evaluation of native and exotic Brazilian plants for anticancer activity. *J. Nat. Med.*, 64, 231-238.

[20] Pereira, F. L., Oliveira, V. B., Viana, C. T. R., Campos, P. P., Silva, M. A. N., Brandão, M. G. L. (2015). Antihyperlipidemic and antihyperglycemic effects of the Brazilian salsaparrilhas *Smilax brasiliensis* Spreng. (*Smilacaceae*) and *Herreria salsaparrilha* Mart. (Agavaceae) in mice treated with a high-refined-carbohydrate containing diet. *Food Res. Int.*, 76, 366-372.

[21] Martins, A. R. (2009). *Morfoanatomia dos órgãos vegetativos e perfil químico de espécies do gênero Smilax L. (Smilacaceae).* [*Morphoanatomy of vegetative organs and chemical profile of species of the genus Smilax L. (Smilacaceae)*]. Thesis (PhD in Plant Biology), Campinas.

[22] Wagner, H., Bladt, S. (1996). *Plant Drug Analysis - a thin layer chromatography atlas.* Berlin: Springer, 2, 73-93.

[23] Amado, P. A. (2016). *Avaliação das atividades antioxidante, alelopática, larvicida e dos efeitos citotóxico e genotóxico das folhas de Smilax brasiliensis Spreng. (Smilacaceae).* [*Evaluation of the*

antioxidant, allelopathic, larvicidal and cytotoxic and genotoxic effects of Smilax brasiliensis Spreng. leaves (Smilacaceae)]. Dissertation (Pharmaceutical Sciences), Divinópolis.

[24] Nunes, A. P. M., Araújo, A. C. (2003). *Ausência de genotoxicidade de esteviosídio em E. coli.* Semana de Iniciação Científica da Universidade do Estado do Rio de Janeiro. [*Absence of stevioside genotoxicity in E. coli.* Scientific Initiation Week of the State University of Rio de Janeiro], 10, 15.

[25] Amado, P. A., Castro, A. H. F., Stein, V. C., Alves, S. N., Lima, L. A. R. S. (2019). Allelopathic, cytotoxic, genotoxic and larvicidal effects of *Smilax brasiliensis* Sprengel leaves. In: *Phytochemicals: Plant Sources and Potential Health Benefits.* New York: Nova Science Publishers.

In: Phytochemicals
Editor: Iman Ryan

ISBN: 978-1-53615-478-8

Chapter 8

ALLELOPATHIC, CYTOTOXIC, GENOTOXIC AND LARVICIDAL EFFECTS OF *SMILAX BRASILIENSIS* SPRENGEL LEAVES

Paula Avelar Amado, Ana Hortência Fonsêca Castro, Vanessa Cristina Stein, Stênio Nunes Alves and Luciana Alves Rodrigues dos Santos Lima
Campus Centro-Oeste Dona Lindu, Universidade Federal de São João Del Rei, Divinópolis, Minas Gerais, Brazil

ABSTRACT

Many commercially available synthetic compounds are used as pesticides and herbicides; however, many of these may be toxic to man and to the environment. Plants may be an alternative source of compounds with similar properties, but without such toxicity. *Smilax brasiliensis* Sprengel is a medicinal species from the Brazilian Cerrado that is used to treat syphilis and skin disorders in folk medicine. This study aimed to investigate the allelopathic, cytotoxic, genotoxic and larvicidal potential of methanol and petroleum ether extracts from *S. brasiliensis* leaves, as well as the effects of their fatty acids and methyl esters. Extracts, fatty acids and methyl esters inhibited the growth of *Allium cepa* radicles and hypocotyls, and pure substances principally inhibited the growth of radicles. For *Lactuca sativa*, samples from *S. brasiliensis* leaves and pure substances showed diverse allelopathic activity with regard to radicle and hypocotyl growth. The frequency of chromosomal abnormalities was significant without a dose-dependent effect. *S. brasiliensis* did not show action on the cell cycle of *A. cepa*. The tested samples showed no larvicidal activity against *Culex quinquefasciatus* larvae. Additional studies should evaluate the possibility of using *S. brasiliensis* extracts, fatty acids and methyl esters as natural herbicides.

Keywords: allelopathy, cerrado, cytotoxicity, medicinal plant, *Smilax*

1. Introduction

In recent years, many scientists around the world have worked toward the purification, characterization and production of bioactive plant substances, leading to the development and industrial use of different products based on native species [1].

Plants produce bioactive substances that differ qualitatively and quantitatively across species and with respect to the place of occurrence of the plant or crop cycle [2]. Secondary metabolites or natural products involved in allelopathy, an ecological phenomenon in which a plant releases chemicals that either have an inhibitory or stimulatory effect on other plants, are called allelochemicals, and are present in tissues from different parts of plants [3].

The production of allelochemicals can be regulated by various environmental factors, such as temperature, light intensity, the availability of water and nutrients, soil texture and the presence of microorganisms [4]. Other factors, such as ultraviolet radiation, disease and insect attacks, can also directly modify the rate of allelochemical production [5]. In addition, factors related to stress may increase the biological activity related to allelochemicals [6, 7].

Allelochemicals can affect seed germination, cell division and differentiation, cell membrane and wall structure and permeability and many other physiological functions [8]. Allelochemicals may retard cell proliferation through inhibition of DNA synthesis [9], disturbances in cytoskeleton formation [10], or by down-regulation of the cell cycle-related genes [11]. Elongation of roots and shoots depends on the normal cell division of meristem cells. Using cytogenetics and other methods, radicle and root growth have shown to be inhibited by different allelochemicals or phytotoxic plant extracts through the disturbance of cell division [9-11].

The allelopathic potential of plants in natural and agricultural management systems may reduce the use of herbicides, insecticides and other pesticides, which would reduce environment and soil pollution as well as autotoxicity hazards [12].

To evaluate substances with allelopathic potential, biological assays are used [13]. In this context, the *A. cepa* test is a fast and sensitive assay that detects environmental genotoxins and mutagens, and also enables the evaluation of the tested agents action mechanisms on the exposed organisms DNA. Thus, this test provides an important method for screening for environmental contamination and results can be extended to other biological systems [14].

Natural products have also been increasingly researched as an option for insect control. Several mosquito species belonging to the *Anopheles*, *Aedes* and *Culex* genera represent key threats to millions of humans and animals worldwide as they play a predominant role in the transmission of malaria, dengue fever, yellow fever, filariasis, chikungunya and Zika [15-20]. *C. quinquefasciatus* is a common and abundant species, occurring predominantly in tropical countries. In humans, it is responsible for the transmission of various diseases, including filariasis, which affects about 146 million people, being is of great importance in public health as it creates administrative and public challenges for the control of mosquito vectors [16].

Many commercially available synthetic compounds are used as pesticides and herbicides. However, many of these products have negative impacts on the environment and are toxic to humans and animals. As such, the use of natural products may be a useful alternative approach due to their less harmful impact on non-target organisms and their biodegradability [21].

Therefore, *S. brasiliensis* (*Smilacaceae*) can be interesting option. This species is a medicinal plant native to the Brazilian Cerrado, popularly known as "salsaparrilha," "japecanga" or "jupicanga" [22-24]. In folk medicine, *S. brasiliensis* is widely used as a diuretic, diaphoretic, stimulant, anti-hypertensive and anti-syphilitic product and is used to treat arthritis, rheumatism and skin disorders [25, 26]. Several studies have reported other pharmacological properties of the *Smilax* genus, such as anticarcinogenic, antioxidant, anti-inflammatory, hypoglycemic, hypotensive, cytotoxic and antifungal activities [27-31]. Steroidal saponins, phytosterols, triterpenoids, flavonoids and phenolic acids are the main components shared by most species of the genus [32-35].

In previous studies, we showed the allelopathic effect of ethanol extract and fractions from *S. brasiliensis* leaves on the growth of *A. cepa* hypocotyls and radicles, and their antioxidant activity [35]. The cytotoxic activity of petroleum ether and methanol extracts, fatty acids and methyl esters from leaves were also evaluated on the larvae of brine shrimp, *Artemia salina* [36]. Nonetheless, there are no reports of *S. brasiliensis* fatty acids and methyl esters, and their allelopathic, cytotoxic and genotoxic effects against test plant organisms. As such, this study aimed to evaluate the allelopathic effects of methanol and petroleum ether extracts from *S. brasiliensis* leaves, as well as their fatty acids and methyl esters, on the growth of *L. sativa* and *A. cepa* seedlings. Their cytotoxic and genotoxic effects on the cell cycle of *A. cepa* and their larvicidal potential against *C. quinque fasciatus* were also investigated.

2. Materials and Methods

2.1. Plant Material

Leaves of *S. brasiliensis* Sprengel were collected in Ijaci South Minas Gerais State, Brazil (21°13'46"S and 44°55'65"W, average altitude 908 m) in October 2014. The plant material was identified by Dr Regina Helena Potsch Andreata and a voucher specimen (57078) was deposited at the Herbarium PAMG of the Company of Agricultural Research of Minas Gerais in Belo Horizonte, MG, Brazil. As *S. brasiliensis* a Brazilian native genetic material, the present study had been approved by the Conselho Nacional de Desenvolvimento Científico e Tecnológico (CNPq) through Access and Shipment Component of Genetic Heritage for scientific research purpose (no. 010455/2014-0) and Certificate of registration for collection of botanical, fungal and microbiological material in SISBIO (no. 24542-5), Register IBAMA (no. 5042260).

2.2. Extraction

Petroleum ether and methanol were used as solvents to obtain the extracts from 300 g of dried leaves using a Soxhlet extractor. The extracts were concentrated in a rotary evaporator at 40°C under reduced pressure to yield petroleum ether (EE, 11.121 g) and methanol (ME, 32.829 g) extracts. Fatty acids and methyl esters were isolated by transesterification reaction, obtained 1.7489 g of fatty acids of EE (FA (E)), 0.045 g of fatty acids of ME (FA (M)), 1.961 g of methyl esters of EE (FAME (E)) and 0.051 g of methyl esters of ME (FAME (M)). Apolar compounds present in the petroleum ether and methanol extracts were identified by gas chromatography/mass spectrometry (GC/MS), in the methyl esters form [36].

2.3. Allelopathic Potential

Germination and seedling growth bioassays are primary tools for determining the allelopathic potential of secondary metabolites (Pawlowski et al. 2013). Allelopathic potential was evaluated on radicle and hypocotyl growth of *L. sativa* (lettuce) var. cabbage (Topsed Garden®, Lot 033242,

Brazil), a dicotyledon species, and *A. cepa* (onion) var. Baia Periforme (Feltrin®, Lot 447395, Brazil), a monocotyledon species [37].

The experiment was carried out in MES [(2-morpholino) ethanesulfonic acid] buffer. Growth studies were conducted in 100-mm Petri containing 25 seeds placed on Whatman no. 1 filter paper. Samples of *S. brasiliensis* and pure substances were applied in various concentrations (250, 500 and 1000 μg/mL per Petri dish) and sealed immediately with Parafilm. A similar treatment with MES served as a control. For each treatment, three replicates were placed in a completely randomized design in a growth chamber and maintained for 10 days in the dark. The temperature was 25°C. After 10 days, the plates were removed from the incubator and chilled at -10°C for 24 h to interrupt the growth process. The effects on the growth can be calculated using the following formula [38]:

$$\% \text{ on the growth } = \frac{\text{Ma}-\text{Mc}}{\text{Mc}} \text{ x } 100 \quad (1)$$

where Ma is the mean value of the seedlings with samples tested and Mc is the mean value for the control (seedlings grown without addition of samples tested). Thus, zero represents the control, positive values represent stimulation of the studied parameter and negative values represent inhibition.

2.4. *Allium cepa* Assay

After 10 days of germination, radicles tips were removed and fixed in Carnoy (ethanol: glacial acetic acid 3:1). Afterwards, radicles were stored at -10 ºC until the time of slide preparation. To prepare the slides, radicles were hydrolyzed in HCl 5 N for 20 min and then washed with distilled water for 10 min. The slides were mounted using the crushing technique [39]. Radicles were then stained with 2% acetic orcein. Slides were analysed in a microscope Primo Stars (Zeiss), coupled with a camera at 400 X magnification. We evaluated three slides per treatment and 350 cells on each slide including 1050 cells per treatment. In addition, the mitotic index (MI) and the total number of chromosomal abnormalities (CA) and necrotic cells (N) were determined at all phases of the cell cycle using the following equations:

$$\mathrm{MI} = \frac{m}{T} \times 100 \tag{2}$$

where *m* is the number of cells in mitosis and T is the total number of cells [40].

$$\mathrm{CA} = \frac{a}{m} \times 100 \tag{3}$$

where *a* is the abnormal cell number and *m* is the number of cells in mitosis [41].

$$\mathrm{MI} = \frac{n}{T} \times 100 \tag{4}$$

where *n* is the number of cells in necrosis and *T* is the total number of cells.

2.5. *Culex quinquefasciatus* Larvicide Assay

The *C. quinquefasciatus* larvae were reared in the Insect Disease Vectors Laboratory at the *Campus* Centro-Oeste Dona Lindu, Federal University of São João Del Rei, according Gerberg [42].

Third and fourth instar larvae of *C. quinquefasciatus* were exposed to different concentrations of the EE, FA (E) and FAME (E) obtained dissolved in dimethyl sulfoxide (DMSO) and water (1:99), to determine the sub-lethal concentrations for seven days. For each sample tested, the larvae were divided into test and control groups, with 60 larvae each, with three replications for each treatment. Negative control was included using distilled water containing the same amount of DMSO as the test sample.

The temperature was maintained at 26 ± 1°C throughout all of the tests. The larvicidal bioassay was performed according to the World Health Organization [43] standard protocols.

2.6. Statistical Analysis

Data obtained were subjected to analysis of variance (ANOVA) followed by the Tukey Test at a 5% probability level. All statistical parameters were calculated using Sisvar software 5.4 [44], the data transformed into square root of X + 0.5, by not Normal distribution. The criterion for statistical significance was set at $p < 0.05$.

3. Results and Discussion

In this study, petroleum ether and methanol extracts obtained from the leaves of *S. brasiliensis* were investigated by gas chromatography-mass spectrometry (GC-MS) to identify major compounds in methyl ester form. The GC-MS analysis revealed unsaturated FAMEs (50.91% EE and 50.73% ME) and saturated FAMEs (49.06% EE and 43.15% ME). In *S. brasiliensis*, methyl linolenate (21.80% EE and 19.04% ME) was the predominant unsaturated FAME, followed by methyl linoleate (18.40% EE and 13.50% ME). Methyl palmitate was the most abundant saturated FAME (29.40% EE and 30.85% ME) [36].

The main fatty acids and FAMEs, which were identified in the extracts through the GC analysis, were subjected to an evaluation of their allelopathic activity and cytogenetic effect for later comparison. The activity of EE, ME, FA (E), FA (M), FAME (E), FAME (M) and pure substances (linoleic acid, linolenic acid, palmitic acid, methyl linoleate, methyl linolenate and methyl palmitate, obtained from Sigma-Aldrich, USA) were evaluated on the radicle and hypocotyl growth of *L. sativa* (lettuce), a dicotyledon species, and *A. cepa* (onion), a monocotyledon species (Figures 1 and 2).

L. sativa seedling growth showed that most 250 µg/mL samples of *S. brasiliensis* significantly inhibited hypocotyl growth, ranging from 72.75% to 92.11%. FA (E) was the only sample that inhibited hypocotyl growth at all concentrations tested. However, at the highest concentration (1000 µg/mL), there was growth stimulus, except for with FA (E) (Figure 1A). At the two lowest concentrations (250 and 500 µg/mL), all samples inhibited radicle growth. The most effective sample was EE at 250 µg/mL, which inhibited growth by 89.46%. At concentrations of 1000 µg/mL, all samples stimulated *L. sativa* radicle growth, with FAME (E) stimulating significant growth (73.15%) compared to the other samples tested (Figure 1B). For the hypocotyl growth of lettuce subjected to pure fatty acid and FAMEs, results were different. Compared with the other samples tested, in the presence of methyl palmitate and methyl linoleate, there was subtle growth at all concentrations, with percentages varying from 2.20% to 12.36% (Figure 1C). In the study of radicle growth, the results were also diverse, especially for palmitic and linoleic acids, which inhibited growth at three concentrations, with percentages varying from 2.35% to 13.48% (Figure 1D).

For *A. cepa* hypocotyl growth, 97.23% inhibition was observed with FA (M), at a concentration of 250 µg/mL, and 52.97% and 75.06% inhibition was observed with FAME (M) at concentrations of 500 and 1000 µg/mL, respectively (Figure 2A). With respect to radicle growth, the most effective were found at 250 µg/mL of FA (M), inhibiting 91.56% of growth. At concentrations of 500 and 1000 µg/mL, the most effective were with FA (E) and FAME (M), inhibiting 73.45% and 74.81% of growth, respectively (Figure 2B). For the pure fatty acids and FAMEs, samples showed inhibitory effects at the lowest concentrations tested, except for palmitic acid, linoleic acid and methyl palmitate (Figure 2C). When evaluating radicle growth, there was a greater predominance of inhibitory potential, except for with linoleic acid, which stimulated growth by 5.68% at the lowest concentration tested (Figure 2D).

The lettuce and onion seed treated with *S. brasiliensis* leaf extracts, fatty acids and FAMEs, in all evaluated concentrations, presented statistically significant differences in growth of hypocotyls and radicles compared to the control groups ($p < 0.0001$).

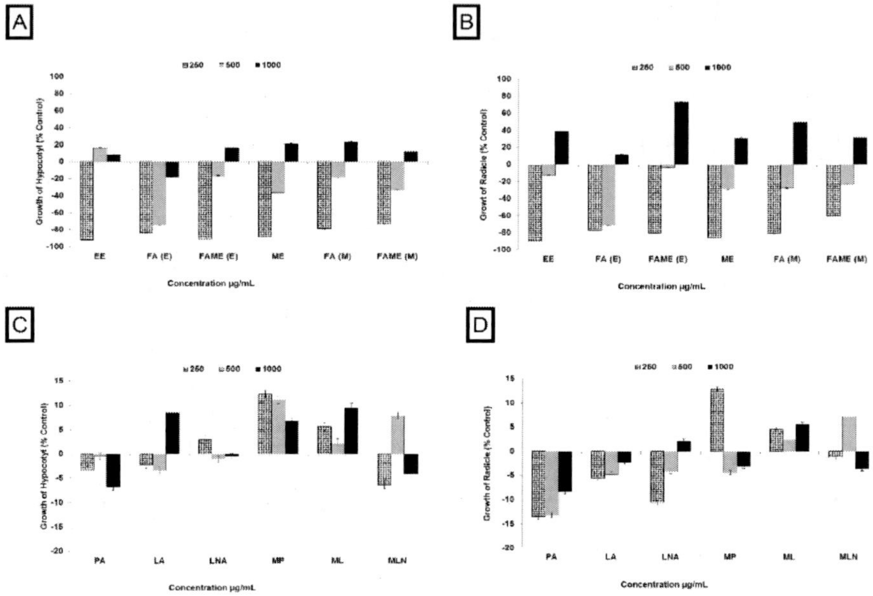

Figure 1. Growth of *L. sativa* seedlings exposed to *S. brasiliensis* extracts, fatty acids and FAMEs and pure fatty acids and FAMEs at 250, 500 and 1000 μg/mL. (A) Hypocotyl and (B) Radicle growth of *L. sativa* exposed to *S. brasiliensis* extracts, fatty acids and FAMEs; (C) Hypocotyl and (D) Radicle growth of *L. sativa* exposed to pure fatty acids and FAMEs. The results are means ± SE (n = 3). Note: Graph scales have been altered for better visualization of results. EE: petroleum ether extract; FA (E): fatty acid of EE; FAME (E): methyl ester of EE; ME: methanol extract; FA (M): fatty acid of ME; FAME (M): methyl ester of ME; PA: palmitic acid; LA: linoleic acid; LNA: linolenic acid; MP: methyl palmitate; ML: methyl linoleate; MLN: methyl linolenate. *There was no statistically significant difference compared to the control group of *L. sativa* hypocotyls with PA (500 μg/mL) and LNA (500 and 1000 μg/mL). For all other samples, there was a statistically significant difference in hypocotyls and radicles compared to the control group ($p < 0.0001$).

For the analyzed standards, there was no statistical difference in lettuce seed hypocotyls with PA (500 μg/mL) and LNA (500 and 1000 μg/mL) compared to the control group. For onion seeds, there was a statistical difference in hypocotyl and radicle compared to the control group ($p < 0.02$). Results demonstrate that *S. brasiliensis* EE, ME, FA (E), FA (M),

FAME (E) and FAME (M) have bioactive substances that interfere with the growth of *L. sativa* and *A. cepa. A. cepa* was more sensitive to extracts, fatty acids and FAMEs.

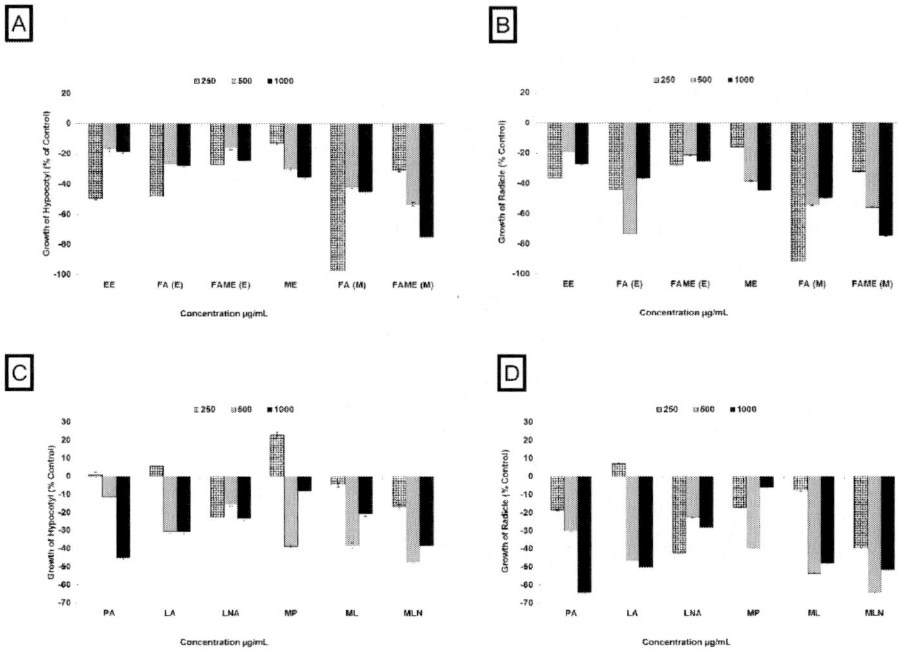

Figure 2. Growth of *A. cepa* seedlings exposed to *S. brasiliensis* extracts, fatty acids and FAMEs and pure fatty acids and FAMEs at 250, 500 and 1000 µg/mL. (A) Hypocotyl and (B) Radicle growth of *A. cepa* exposed to *S. brasiliensis* extracts, fatty acids and FAMEs; (C) Hypocotyl and (D) Radicle growth of *A. cepa* exposed to pure fatty acids and FAMEs. The results are means ± SE (n = 3). Note: Graph scales have been altered for better visualization of results. EE: petroleum ether extract; FA (E): fatty acid of EE; FAME (E): methyl ester of EE; ME: methanol extract; FA (M): fatty acid of ME; FAME (M): methyl ester of ME; PA: palmitic acid; LA: linoleic acid; LNA: linolenic acid; MP: methyl palmitate; ML: methyl linoleate; MLN: methyl linolenate. *All samples presented statistically significant differences in hypocotyls and radicles compared to the control group ($p < 0.02$).

Results also show that radicle and hypocotyl growth cannot be explained by dose dependence. When comparing the results of the effects of fatty acids and FAMEs from *S. brasiliensis* leaves to pure substances, fatty acids and FAMEs showed a higher efficacy than pure substances; this suggests the possibility of synergism. According to Odum [45], synergism is a chemical phenomenon in which the effect obtained by the combined action of two different chemicals is greater than the sum of the individual effects of those substances. Thus, while the pure substances have some potential to act as herbicides alone, this property may be elevated by synergism with other compounds present in the samples. In the literature, there is some evidence to suggest that synergism may exist as a result of combinatorial action between different allelochemicals [46-49].

Studies on allelopathic activity for the *Smilax* genus are still rare; only one study reported an allelopathic potential of the ethanol extract and the hexane and dichloromethane fractions from *S. brasiliensis* leaves, with inhibition greater than 60% for the hypocotyl and radicle growth of onion seedlings, and varied results for lettuce seeds [35]. Reports of allelopathic effects of fatty acids and FAMEs are also rare. Pinto et al. [38] demonstrated an allelopathic potential of FAMEs obtained from corn and sunflower oils, with a high percentage of inhibition of *A. cepa* radicle and hypocotyl growth. Fatty acids and FAMEs of ripe and unripe *Solanum lycocarpum* fruits exhibited allelopathic effects on *L. sativa* and *A. cepa* seeds, with inhibition of hypocotyl and radicle growth [50].

Chemical compounds which have allelopathic effects can also be genotoxic and mutagenic [51]. Allelochemicals may inhibit germination by altering membrane permeability, respiration, enzyme activities, DNA transcription and RNA translation, or alternatively, by a combination of these factors [52]. The MI in *A. cepa* meristematic cells has been reported to be an efficient indicator of direct action on DNA [14]; this motivated the MI analysis of onion meristematic cells subjected to different samples of *S. brasiliensis*, and pure substances were employed for cytotoxicity evaluation. Significant changes were observed in the MI with FA (M) and FAME (E) (Figure 3A). The MI decreased with an increase in EE concentration; however, with FA (E), MI was affected only at the highest concentration. With FA (M), at a concentration of 250 μg/mL, the MI was stimulated and inhibited at 500 μg/mL, consistent with the inhibition observed in radicle growth (Figure 2B). For pure substances, stimulation was observed in the MI at concentrations of 250 and 1000 μg/mL of linoleic acid (Figure 3B). The stimulation of radicle growth at the lowest tested concentration of linoleic acid (Figure 2D) may be due to stimulation of cell division.

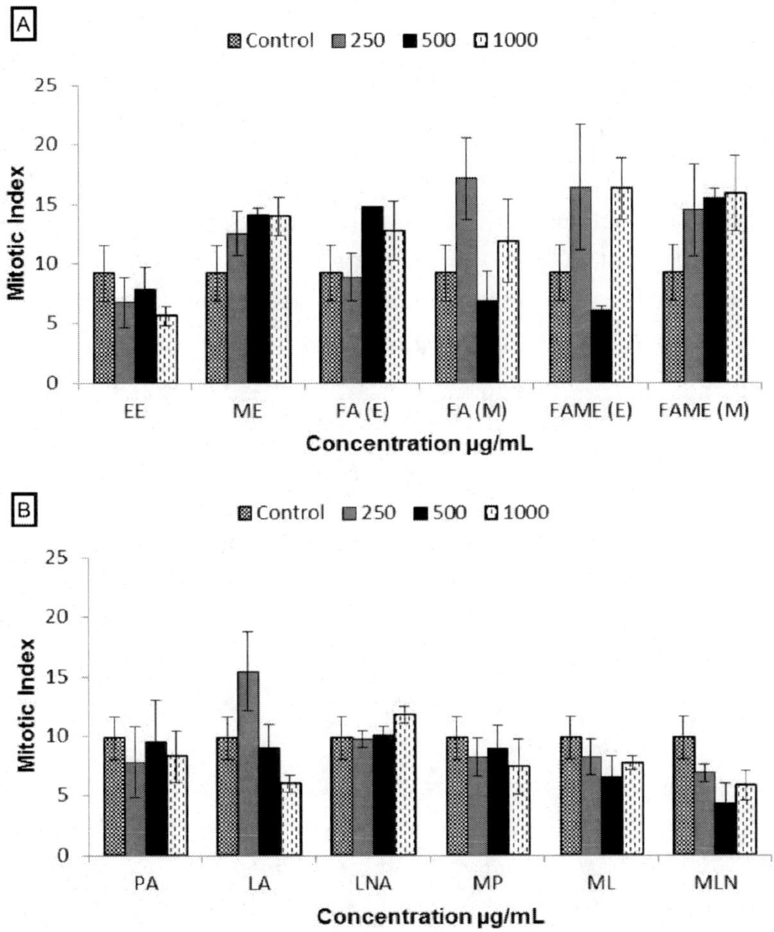

Figure 3. Mitotic index of *A. cepa* cells treated with (A) *S. brasiliensis* extracts, fatty acids and FAMEs and (B) pure fatty acids and FAMEs. The results are means ± SE (n = 3). EE: petroleum ether extract; FA (E): fatty acid of EE; FAME (E): methyl ester of EE; ME: methanol extract; FA (M): fatty acid of ME; FAME (M): methyl ester of ME; PA: palmitic acid; LA: linoleic acid; LNA: linolenic acid; MP: methyl palmitate; ML: methyl linoleate; MLN: methyl linolenate. *For all samples, there were no statistically significant differences compared to the control.

Necrotic cells were also observed. However, there was no significant difference between the substances tested. The appearance of abnormal seedlings is one of the effects of allelopathic substances, and necrosis of radicles is one of the most common symptoms [51]. This process leads to metabolic damage and, finally, to death of the cell; it can be of several types: spontaneous; triggered by oxidative or stress mechanisms; or induced by infectious agents, toxic chemicals and hypersensitivity [53].

Analysis of cell division revealed the presence of micronuclei, chromosomal fragment loss, chromosomal bridges, sticky chromosomes and abnormalities in the mitotic spindle, including cells with irregular metaphase, irregular anaphase and vagrant chromosomes (Figure 4); however statistical analysis did not show a significant difference between the tested groups. As was observed in our work, Forskolin, isolated from *Coleus forskohlii* roots, showed a reduction in MI at certain concentrations, as well as the presence of several chromosomal aberrations, such as sticky metaphases, sticky anaphases, micronuclei, chromosomal breaks and chromosomal fragments [54]. Chromosomal abnormalities were also observed in all stages of mitosis with all treatments (Figure 4). These changes may also indirectly lead to modifications in the genetic material. The cytotoxic effect is not necessarily associated with a genotoxic effect [55], which explains how genotoxic changes were not detected in the meristematic cells treated with samples and pure substances, even though there was a change in the MI.

Greater MIs than those seen in the negative control may indicate the induction of increased cell division, which characterizes cell damage. Smaller MIs than those seen in the negative control might indicate that the growth and development of organisms are being affected by components in the samples [56]. Plant extracts may have cytotoxic effects due to inhibition of the cell cycle [57]. DNA damage can induce genotoxicity in plant cells and affect cell mitosis; this poses one of the major threats to radicle growth since radicle growth depends on cell mitosis [58].

Studies evaluating the cytogenetic effect of extracts using the *A. cepa* test are uncommon.

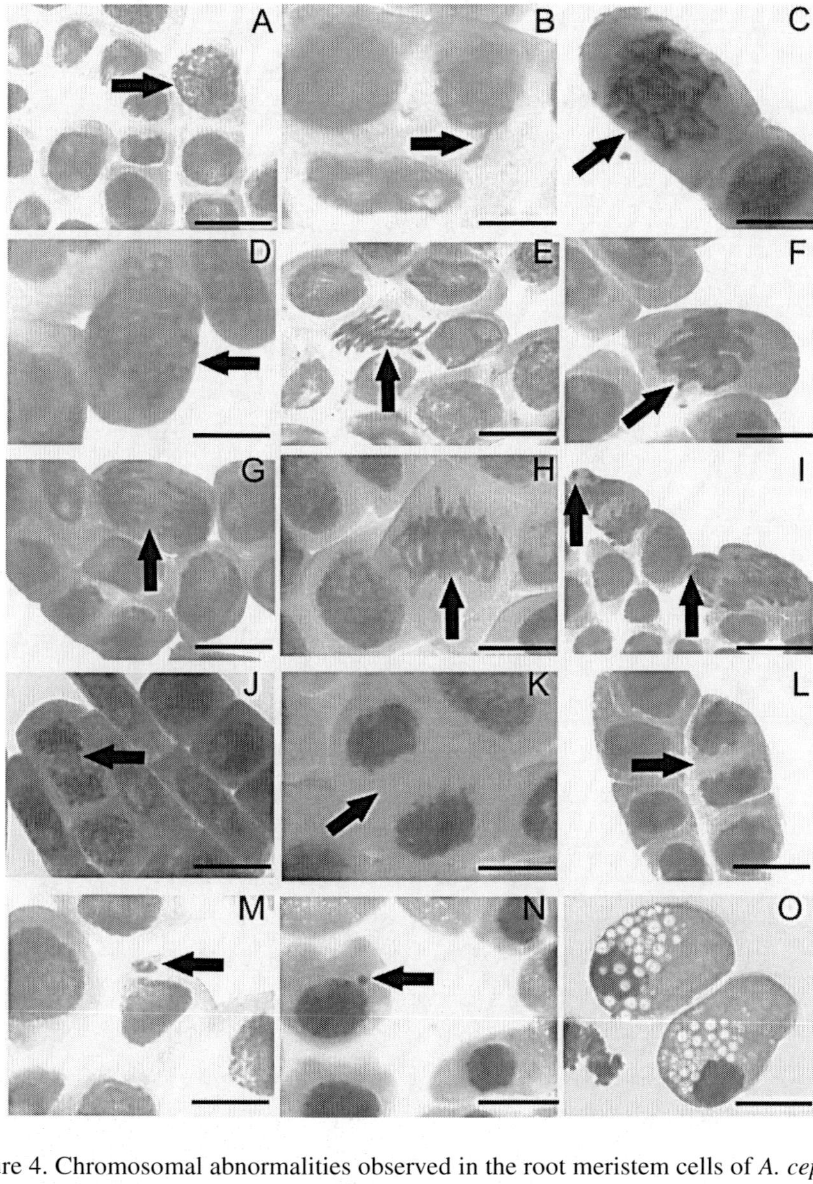

Figure 4. Chromosomal abnormalities observed in the root meristem cells of *A. cepa* after treatment with EE, EM, fatty acids and FAMEs of *S. brasiliensis*. (A) normal prophase; (B) prophase with vagrant chromosome; (C–D) irregular prophase; (E–F) irregular metaphase; (G) normal anaphase; (H) irregular anaphase; (I) anaphase with chromosomal losses and vagrant chromosome; (J) anaphase with chromosomal bridge; (K) normal telophase; (L) telophase with sticky chromosomes; (M–N) micronucleus; (O) necrosis. Bar 10 μm.

Iganci et al. [59] showed that the aqueous extract of different species of boldo (*Plectranthus barbatus*, *Plectranthus amboinicus* and *Vernonia condensata*) produced changes in the process of cell division in onions, but cytotoxic and genotoxic effects were not detected; however, there was a significant increase in the cell division rate relative to the control. Another study of aqueous extracts of neem (*Azadirachta indica*), eucalyptus (*Eucalyptus globulus*) and thyme (*Thymus vulgaris*) concluded that all of the tested aqueous extracts caused cytotoxic and genotoxic damage [60]. Silva et al. [61] studied the aqueous extract of *Erythrina velutina,* and noted that it could have genotoxic effects at some doses. Studies of cytotoxicity and genotoxicity with fatty acid and FAMEs are yet to be performed.

An alternative and recent approach to mosquito control is the use of natural plant products. Many natural products are effective, environment friendly, easily biodegradable and inexpensive. They have no negative effects on non-target organisms and have different innovative modes of action [62].

To determine the mortality rate of *C. quinquefasciatus*, larvae were observed until adulthood. EE, FA (E) and FAME (E) showed no larvicidal activity on *C. quinquefasciatus* larvae at any tested concentrations. It was not possible to calculate LC_{50} and LC_{90} values because of low larval lethality. Previous studies have demonstrated larvicidal activity for the acids that were identified in this study, such as oleic, linoleic and linolenic acid, with LC_{50} ranging from 8.58 to 190.78 mg/L [63]. These same acids, tested by Melo et al. [63], have been reported to be toxic against *Culex pipiens pallens*, *Aedes aegypti*, and *Aedes albopictus* [64]. Data from the literature suggest that fatty acids may be most selective due to the chain length or positions of the double bonds, since unsaturated fatty acids were more toxic than saturated fatty acids [63, 64]. There are no previous reports of the larvicidal effects of *S. brasiliensis* against *C. quinquefasciatus*. For the *Smilax* genus, only a single study demonstrated the larvicidal activity of *S. larvata* ethanol extract on *Aedes aegypti* larvae, with LC_{50} = 225 μg/mL and LC_{90} = 350 μg/mL [65].

Conclusion

In summary, results show that the extracts, fatty acids and methyl esters of *S. brasiliensis* leaves exhibited significant allelopathic effects. The allelopathic effect on *A. cepa* seedlings showed promising results, predominantly with growth inhibition of hypocotyls and radicles at all tested concentrations. The tested samples showed no larvicidal activity against the *C. quinquefasciatus* mosquito. Further studies should evaluate the possibility of using *S. brasiliensis* extracts, fatty acids and FAMEs as natural sources of herbicides, since they pose no risk to nature, or even to humans, with no demonstrated genotoxic effects.

References

[1] El-Gawad, A. A. M. (2016). Chemical constituents, antioxidant and potential allelopathic effect of the essential oil from the aerial parts of *Cullen plicata*. *Ind. Crops Prod.*, 80, 36-41.

[2] Cherdshewasart, W., Subtang, S., Dagkab, W. (2007). Major isoflavonoid contents of the phytoestrogen rich-herb *Pueraria mirifica* in comparison to *Pueraria lobata*. *J. Pharm. Biomed. Anal.*, 43, 428-434.

[3] Cheng, F., Cheng, Z. (2015). Research progress on the use of plant allelopathy in agriculture and the physiological and ecological mechanisms of allelopathy. *Front. Plant Sci.*, 6, 1-16.

[4] Macías, F. A., Molinillo, J. M. G., Varela, R. M., Galindo, J. C. G. (2007). Allelopathy - a natural alternative for weed control. *Pest. Manag. Sci.*, 63, 327-348.

[5] Formagio, A. S. N., Masetto, T. E., Baldivia, D. S., Vieira, M. C., Zarate, N. A. H., Pereira, Z. V. (2010). Allelopathic potential of five species of the Annonaceae family. *R. Bras. Bioci.*, 8, 349-354.

[6] Tongma, S., Kobayashi, K., Usui, K. (2001). Allelopathic activity of Mexican sunflower [*Tithonia diversifolia* (Hemsl.) A. Gray] in soil

under natural field conditions and different moisture conditions. *Weed Biol. Manag.*, 1, 115-119.

[7] Povh, J. A., Pinto, D. D., Corrêa, M. O. G., Ono, E. O. (2007). Atividade alelopática de *Machaerium acutifolium* Vog. na germinação de *Lactuca sativa* L. [Allelopathic activity of *Machaerium acutifolium* Vog. on the germination of *Lactuca sativa* L.] *R. Bras. Bioci.*, 5, 447-449.

[8] Chon, S. U., Nelson, C. J. (2010). Allelopathy in Compositae plants. *Agron. Sust. Dev.*, 30, 349-358.

[9] Nishida, N., Tamotsu, S., Nagata, N., Saito, C., Sakai, A. (2005). Allelopathic effects of volatile monoterpenoids produced by *Salvia leucophylla*: inhibition of cell proliferation and DNA synthesis in the root apical meristem of *Brassica campestris* seedlings. *J. Chem. Ecol.*, 31, 1187-1203.

[10] Soltys, D., Rudzinska-Langwald, A., Kurek, W., Gniazdowska, A., Sliwinska, E., Bogatek, R. (2011). Cyanamide mode of action during inhibition of onion (*Allium cepa* L.) root growth involves disturbances in cell division and cytoskeleton formation. *Planta*, 234, 609-621.

[11] Zhang, Y., Gu, M., Shi, K., Zhou, Y. H., Yu, J. Q. (2010). Effects of aqueous root extracts and hydrophobic root exudates of cucumber (*Cucumis sativus* L.) on nuclei DNA content and expression of cell cycle-related genes in cucumber radicles. *Plant Soil*, 327, 455-463.

[12] Anaya, A. L. (2006). Allelopathic organisms and molecules: promising bioregulators for the control of plant diseases, weeds, and others pests. In: *Allelochemicals: Biological Control of Plant Pathogens and Diseases.* Dordrecht: Kluwer Academic Publishers, 31-78.

[13] Souza, S. A. M., Stein, V. C., Cattelan, L. V., Bobrowski, V. L., Rocha, B. H. G. (2005). Utilização de sementes de alface e rúcula como ensaios biológicos para avaliação do efeito citotóxico e alelopático de extratos aquosos de plantas medicinais. [Use of lettuce and arugula seeds as biological assays to evaluate the cytotoxic and

allelopathic effect of aqueous medicinal plants extracts]. *Rev. Biol. Ciênc. Terra*, 5, 3-9.

[14] Leme, D. M., Marin-Morales, M. A. (2009). *Allium cepa* test in environmental monitoring: A review on its application. *Mutat. Res. - Rev. Mutat. Res.*, 682, 71-81.

[15] Jiang, J., Xu, Q. (2003). Immunomodulatory activity of aqueous extract from rhizome of *Smilax glabra* in the later phase of adjuvant-induced arthritis in rats. *J. Etnhopharmacol.*, 85, 53-59.

[16] Elumalai, D., Kaleena, P. K., Fathima, M., Muttapan, M. (2013). Evaluation of biological activity of *Hyptis suaveolens* (L) Poit and *Leucas aspera* (Willd) against *Culex quinquefasciatus. Int. J. Biosci. Res.*, 2, 1-6.

[17] Baden, L. R., Petersen, L. R., Jamieson, D. J., Powers, A. M., Honein, M. A. (2016). Zika virus. *N. Engl. J. Med.*, 374, 1552-1563.

[18] Benelli, G., Mehlhorn, H. (2016). Declining malaria, rising of dengue and Zika virus: insights for mosquito vector control. *Parasitol. Res.*, 115, 1747-1754.

[19] Moulin, E., Selby, K., Cherpillod, P., Kaiser, L., Boillat-Blanco, N. (2016). Simultaneous outbreaks of dengue, chikungunya and Zika virus infections: diagnosis challenge in a returning traveler with nonspecific febrile illness. *New Microb. New Infect.*, 11, 6-7.

[20] Mayer, S. V., Tesh, R. B., Vasilakis, N. (2017). The emergence of arthropod-borne viral diseases: A global prospective on dengue, chikungunya and Zika fevers. *Acta Trop.*, 166, 155-163.

[21] Prabakar, K., Jebanesan, A. (2004). Larvicidal efficacy of some Cucurbitacious plant leaf extracts against *Culex quinquefasciatus* (Say). *Bioresour. Technol.*, 95, 113-114.

[22] Pio Corrêa, M. (1969). *Dicionário da plantas úteis do Brasil e das exóticas cultivadas.* [*Dictionary of useful plants of Brazil and exotic cultivated*]. Rio de Janeiro.

[23] Guarim Neto, G., Morais, G. (2003). Medicinal plants resources in the Cerrado of Mato Grosso State, Brazil: a review. *Acta Bot. Bras.*, 17, 561-584.

[24] Medeiros, M. F., Valle, L. S., Andreata, R. H. P. (2007). Histórico e o uso de "salsaparrilha" (*Smilax* spp.) pelos boticários no Mosteiro de São Bento. [History and the use of "sarsaparilla" (*Smilax* spp.) by apothecaries in the Monastery of São Bento]. *R. Bras. Bioci.*, 5, 27-29.

[25] Lorenzi, H., Matos, F. J. A. (2008). *Plantas medicinais no Brasil: nativas e exóticas.* [*Medicinal plants in Brazil: native and exotic*]. São Paulo: Nova Odessa.

[26] Breitbach, U. B., Niehues, M., Lopes, N. P., Faria, J. E. Q., Brandão, M. G. L. (2013). Amazonian Brazilian medicinal plants described by C. F. P. von Martius in the 19th century. *J. Ethnopharmacol.*, 147, 180-189.

[27] Xia, D., Yu, X., Liao, S., Shao, Q., Mou, H., Ma, W. (2010). Protective effect of *Smilax glabra* extract against lead-induced oxidative stress in rats. *J. Ethnopharmacol.*, 130, 414-420.

[28] Amaro, C. A. B., González-Cortazar, M., Herrera-Ruiz, M., Román-Ramos, R., Aguilar-Santamaría, L., Tortoriello, J., Jiménez-Ferrer, E. (2014). Hypoglycemic and hypotensive activity of a root extract of *Smilax aristolochiifolia,* standardized on N-trans-feruloyl-tyramine. *Molecules*, 19, 11366-11384.

[29] Morais, M. I., Pinto, M. E. A., Araújo, S. G., Castro, A. H. F., Duarte-Almeida, J. M., Rosa, L. H., Rosa, C. A., Johann, S., Lima, L. A. R. S. (2014). Antioxidant and antifungal activities of *Smilax campestris* Griseb. (*Smilacaceae*). *Nat. Prod. Res.*, 28, 1275-1279.

[30] Liang, F., He, J. W., Zhu, G. H., Liu, R. H., Shu, J. C., Shao, F., Huang, H. L. (2016). New steroidal saponins with cytotoxic activities from *Smilax trinervula*. *Phytochem. Lett.*, 16, 294-298.

[31] Yaacob, W. M. H. W., Ahmad, N., Sulain, M. D. (2016). Evaluations of cytotoxicity of *Smilax myosotiflora* and its effects on sexual hormone levels and testicular histology in male rats. *Asian Pac. J. Trop. Biomed.*, 6, 246-250.

[32] Xu, J., Li, X., Zhang, P., Zhan, L. L., Wang, Y. (2005). Antiinflammatory constituents from the roots of *Smilax bockii* Warb. *Arch. Pharm. Res.*, 28, 395-399.

[33] Cáceres, A., Cruz, S. M., Martínez, V., Gaitán, I., Santizo, A., Gattuso, S., Gattuso, M. (2012). Ethnobotanical, pharmacognostical, pharmacological and phytochemical studies on *Smilax domingensis* in Guatemala. *Rev. Bras. Farmacogn.*, 22, 239-248.

[34] Wu, X. H., Zhang, J., Wang, S. Q., Yang, V. C., Anderson, S., Zhang, Y. W. (2014). Riparoside B and timosaponin J, two steroidal glycosides from *Smilax riparia*, resist to hyperuricemia based on URAT1 in hyperuricemic mice. *Phytomedicine*, 21, 1196-1201.

[35] Fonseca, J. C., Barbosa, M. A., Silva, I. C. A., Duarte-Almeida, J. M., Castro, A. H. F., Dos Santos Lima, L. A. R. (2017). Antioxidant and allelopathic activities of *Smilax brasiliensis* Sprengel (*Smilacaceae*). *S. Afr. J. Bot.*, 111, 336-340.

[36] Amado, P. A., Ferraz, V., Silva, D. B., Carollo, C. A., Castro, A. H. F., Lima, L. A. R. S. (2018). Chemical composition, antioxidant and cytotoxic activities of extracts from the leaves of *Smilax brasiliensis* Sprengel (*Smilacaceae*). *Nat. Prod. Res.*, 32, 610-615.

[37] Freitas, L. B. O., Boaventura, M. A. D., Santos, W. L., Stehmann, J. R., Junior, D. D., Lopes, M. T. P., Magalhães, T. F. F., Silva, D. L., Resende, M. A. (2015). Allelopathic, cytotoxic and antifungic activities of new dihydrophenanthrenes and other constituents of leaves and root extracts of *Banisteriospsis anisandra* (Malpighiaceae). *Phytochem. Lett.*, 12, 9-16.

[38] Pinto, M. E. A., Silva, N. L., Siqueira, E. P., Santos, F. J. L., Lima, L. A. R. S. (2013). Allelopathic effect of fatty acid methyl esters from corn and sunflower oils. *BBR - Biochem. Biotechnol. Rep.*, 2, 44-48.

[39] Guerra, M., Souza, M. J. (2002). *Como observar cromossomos: um guia de técnicas em citogenética vegetal, animal e humana*. [*How to observe chromosomes: a guide to techniques in plant, animal and human cytogenetics*]. São Paulo: Funpec.

[40] Pires, N. M., Souza, I. R. P., Prates, H. T., Faria, T. C. L., Filho, I. A. P., Magalhães, P. C. (2001). Effect of leucaena aqueous extract on the development, mitotic index and peroxidase activity in maize seedlings. *Rev. Bras. Fisiol. Veg.*, 13, 55-65.

[41] Rãcuciu, M. (2009). Effects of radiofrequency radiation on root tip cells *Zea mays. Rom. Biotechnol. Lett.*, 14, 4365-4369.

[42] Gerberg, E. J. (1980). Manual for mosquito rearing and experimental techniques. *Am. Mosq. Control Assoc.*, 5, 1-124.

[43] World Health Organization. (1981). *Instructions for determining the susceptibility or resistance of adult mosquitoes to organochlorine, organophosphorous and carbamate insecticides.* Geneva, (WHO/ VBC81.805).

[44] Ferreira, D. F. (2011). Sisvar: a computer statistical analysis system. *Ciênc. Agrotec.*, 35, 1039-1042.

[45] Odum, E. (2007). *Fundamentos de Ecologia.* [*Fundamentals of Ecology*]. São Paulo: Thomson.

[46] Weidenhamer, J. D. (1994). Allelopathic potential of menthofuran monoterpenes from *Calamintha ashei. J. Chem. Ecol.*, 20, 3345-3359.

[47] Einhellig, F. A. (1995). Mechanism of action of allelochemicals in allelopathy. In: Inderjit, K. M. M. D., Einhellig, F. A. (eds.). *Allelopathy: Organisms, processes, and applications.* Washington DC: American Chemical Society, 96-116.

[48] Ferreira, A. G. (2005). Alelopatia: sinergismo e inibição. [Allelopathy: synergism and inhibition]. In: Nogueira, R. J. M. C., Araújo, E. L., Willadino, L. G., Cavalcante, U. M. T. (eds.). *Estresses ambientais: danos e benefícios em plantas.* [*Environmental stresses: damages and benefits in plants*]. Recife: Imprensa Universitária, 433-440.

[49] Souza Filho, A. P. S., Borges, F. C., Santos, L. S. (2006). Comparative analysis of the allelopathic effects of the chemical compounds tithonine and acetylated tithonine. *Planta Daninha*, 24, 205-210.

[50] Silva, V. C. B., Amado, P. A., Siqueira, E. P., Alves, S. N., Duarte-Almeida, J. M., Lima, L. A. R. S. (2018). Antioxidant and Allelopathic Potential of Oils, Fatty Acids, and Methyl Esters from Fruits of *Solanum lycocarpum* A. St. Hil. (Solanaceae). In:

Solanaceae: Cultivation, Nutrition and Health. New York: Nova Science Publishers, 79-94.

[51] Ferreira, A. G., Áquila, M. E. A. (2000). Alellopathy: an emerging topic in ecophysiology. *Rev. Bras. Fisiol. Veg.*, 12, 175-204.

[52] El-Shora, H. M., El-Gawad, A. A. M. (2014). Evaluation of allelopathic potential of *Rumex dentatus* root extract and allelochemicals on *Cicer arietinum*. *J. Stress Physiol. Biochem.*, 10, 167-180.

[53] Zakeri, Z., Lockshin, R. A. (2002). Cell death during development. *J. Immunol. Methods*, 265, 3-20.

[54] Mohammed, K. P., Aarey, A., Tamkeen, S., Jahan, P. (2015). Forskolin: Genotoxicity assessment in *Allium cepa*. *Mutat. Res. Toxicol. Environ. Mutagen.*, 777, 29-32.

[55] Oliveira, L. M., Voltolini, J. C., Barbério, A. (2011). Mutagenic potential of pollutants in the water of the Paraíba do Sul river, Tremembé, SP, Brazil, using the *Allium cepa* test. *Rev. Ambient. Água - An Interdisciplinary Journal of Applied Science*, 6, 90-103.

[56] Caritá, R., Marin-Morales, M. A. (2008). Induction of chromosome aberrations in the *Allium cepa* test system caused by exposure of seeds to industrial effluents contaminated with azo dyes. *Chemosphere*, 72, 722-725.

[57] Çelik, T. A., Äslanturk, Ö. S. (2006). Anti-mitotic and anti-genotoxic effects of *Plantago lanceolata* aqueous extracts on *Allium cepa* root tip meristem cells. *Biologia (Bratisl.)*, 61, 693-697.

[58] Qin, R., Wang, C., Chen, D., Björn, L. O., Li, S. (2015). Copper-induced root growth inhibition of *Allium cepa* var. *agrogarum* L. involves disturbances in cell division and DNA damage. *Environ. Toxicol. Chem.*, 34, 1045-1055.

[59] Iganci, J. R. V., Bobrowski, V. L., Heiden, G., Stein, V. C., Rocha, B. H. G. (2006). Effect of aqueous extracts of the "boldo" species on the germination and mitotic index of *Allium cepa* L. *Arq. Inst. Biol.*, 73, 79-82.

[60] Qureshi, S. T., Chandio, P., Noman, A., Parveen, A., Soomro, Y. (2015). Cytotoxic and genotoxic and oxidative effects of aqueous

extracts of some frequently used medicinal plants in Pakistan. *J. Bot. Res. Inst. Tex.*, 8, 29-35.

[61] Silva, D. S. B. S., Anuska, C. F. S., Garcia, S. S., Mata, B. O., Charles, S. E., Scher, R., Pantaleao, S. M. (2011). Genotoxicity and cytotoxicity of *Erythrina velutina* Willd., Fabaceae, on the root meristem cells of *Allium cepa* L., Amaryllidaceae. *Braz. J. Pharmacog.*, 21, 92-97.

[62] George, D. R., Finn, R. D., Graham, K. M., Sparagano, O. A. (2014). Present and future potential of plant-derived products to control arthropods of veterinary and medical significance. *Parasit. Vectors*, 7, 1-12.

[63] Melo, A. R., Garcia, I. J. P., Serrão, J. E., Santos, H. L., Lima, L. A. R. S., Alves, S. N. (2018). Toxicity of different fatty acids and methyl esters on *Culex quinquefasciatus* larvae. *Ecotoxicol. Environ. Saf.*, 154, 1-5.

[64] Perumalsamy, H., Jang, M. J., Kim, J. R., Kadarkari, M., Ahnn, I. J. (2015). Larvicidal activity and possible mode of action of four flavonoids and two fatty acids identified in *Millettia pinnata* seed toward three mosquito species. *Parasit. Vectors*, 8, 237.

[65] Hirota, B. C. K., Merino, F. J. Z., Verdam, M. C. S., Silva, C. B., Murakami, F. S., Miguel, O. G., Miguel, M. D. (2015). Larvicide and antifungal activities of sarsaparilla (*Smilax larvata*) extracts. *Int. J. Pharm. Pharm. Sci.,* 7, 308-311.

In: Phytochemicals
Editor: Iman Ryan

ISBN: 978-1-53615-478-8

Chapter 9

BIOLOGICAL POTENTIAL OF THE SPECIES *TECOMA STANS* (L.) KUNTH (*BIGNONIACEAE*)

Thaís Paula Rodrigues Gonçalves, Paula Avelar Amado, Iara Cotta Mendes, Adriano Guimarães Parreira and Luciana Alves Rodrigues dos Santos Lima

Campus Centro-Oeste Dona Lindu, Federal University of São João Del Rei, Divinópolis, Minas Gerais, Brazil

ABSTRACT

Nowadays there is increasing interest in medicinal plants as well as the use of herbal medicines to treat various pathologies. The species *Tecoma stans* (L.) Kunth belongs to the *Bignoniaceae* family which is composed of 860 species, grouped in 100 predominantly Neotropical genera. *T. stans* is a tree native to Mexico and the southern United States, having been introduced in Brazil as an ornamental plant. It occurs as a cultivated plant or of sub-spontaneous origin in almost the entire territory. This literature review synthesizes phytochemical aspects recently elucidated and biological activity for *T. stans* species.

Phytochemical investigations have revealed the presence of saponins, flavonoids, monoterpene alkaloids, tannins, phenolic compounds, steroids, anthraquinones, terpenes, hydrocarbons, volatile oils, glycosides, carbohydrates, proteins, and resins in *T. stans* extracts. Throughout the world, several types of biological activity have been reported for the species, including antimicrobial, antioxidant, antidiabetic, antitumor, antinociceptive, anti-inflammatory, and cytotoxic activity. However, as it is considered exotic in Brazil and is used on a large scale for ornamentation, studies of the medicinal potential of *T. stans* in the country are still rare.

The variety of biological activity together with the phytochemical composition of *T. stans* suggest that the species represents an important source of medicinal properties and has great potential for improvement of the quality of life and welfare of the population, evidencing the need for and relevance of more studies of this species, especially in Brazil.

Keywords: medicinal plants, *Tecoma stans*, extracts, phytochemistry, herbal medicine

1. THE FAMILY *BIGNONIACEAE* AND THE *TECOMA* GENUS

The *Bignoniaceae* family is composed of seven tribes with approximately 100 genera and 860 species [1], distributed mainly in the tropical and subtropical regions of the planet, being especially diverse in South America [2]. This family is inserted in the order Lamiales, being closely related to the families Verbenaceae and Acanthaceae [3]. It consists of shrubs, trees, and climbing plants. Species belonging to this family have great medicinal potential, and since 1880, studies have already described their therapeutic potential [4].

The economic importance of this family is associated, mainly, with the potential for logging, furniture manufacturing, civil construction, and handicrafts. Some species, however, are used in pharmacology, and many representatives of this family are ornamental, due to the beauty of their flowers. Among these, some have been selected by countries as their national symbol, including El Salvador (*Tabebuia rosea* (Bertol.) DC.), Venezuela (*Handroanthus bilbergii* (Bureau & K. Schum.) S. Grose),

Brazil (*H. serratifolius* (Vahl) S. Grose), Ecuador (*H. chrysanthus* (Jacq.) S. Grose), Paraguay (*Tabebuia* sp.), and Argentina (*Jacaranda mimosifolia* D. Don) [5].

The *Tecoma* genus has 209 species cataloged, with 203 native species, five cultivated and one naturalized, 11 subspecies and seven varieties in the Neotropical zone [6]. Reports reveal that several bioactive compounds have been identified in specimens of the *Tecoma* genus, such as saponins, flavonoids, alkaloids, phenols, steroids, anthraquinones, tannins, terpenes, hydrocarbons, volatile oils and glycosides, exhibiting antioxidant, antibacterial, and antifungal activity [7-9].

2. The *Tecoma stans* Species

Tecoma stans (L.) Kunth (*Bignoniaceae*) is a tree native to Mexico and the southern United States [10]. It presents vigorous and rapid growth with intense seed production besides efficient caulinar and radicular budding through gemiferous roots [11]. Its leaves are composed of oval-lanceolate, sub-sessile and serrated leaflets. The inflorescences are terminal or axillary, with many tubular, yellow flowers. The fruits of *T. stans* are dried dehiscents of the pod type, approximately 12 cm long, which are filled with small winged seeds. The seeds are exalbuminous, with a membranous edge, composed of cells with thick walls and cellulosic material [12].

T. stans (Figure 1) has three varieties: stans, angustata and velutina, but in Brazil only the stans variety occurs. It is not known when the species was introduced into the country; the oldest references to it is dated 1871, being used as an ornamental plant [13]. The species occurs as cultivated or sub-spontaneous in almost all the Brazilian territory, which gives it an ecologically large dimension. Exotic plants are usually grown in urban areas, gardens or backyards of residences, presenting diverse functions, from the production of fruits to medicinal use, shading and ornamentation, due to the beauty of the flowers. They also play an important role in the formation of biological corridors that connect nearby forest fragments and provide resources for pollinators [14, 15]. In northern Paraná, this

Bignoniaceae species is cited as causing damage to pastures and the environment. *T. stans* has invaded about 50,000 hectares of pasture of this region and is in a frank process of expansion [16], being mentioned as one of the main invasive species in Brazil.

Today, it is known that *T. stans* has various medicinal and therapeutic properties. The leaves of *T. stans* exhibit anti-inflammatory and analgesic potential [17] and have been evaluated as antibacterial agents against some human bacterial strains [18]. The leaf infusion has diuretic properties and is used in diabetes treatment, as well as for intestinal and stomach problems [19]. In addition, *T. stans* trees provide firewood and charcoal, and their wood is used in building construction [20]. *T. stans* also disrupts the anticoagulant effect of a reference drug [21]. Extracts from the branches and leaves exhibit medicinal properties [22]. However, studies on this species are still necessary, especially in Brazil.

Figure 1. The species *Tecoma stans* (L.) Kunth. A) tree, B) flowers.

2.1. Phytochemical Constituents of *T. stans*

The literature reveals that *T. stans* possesses various well-known bioactive compounds. The biological activity reported for this species may be related to the presence of one or more of its active constituents [23].

Studies have revealed the presence of saponins, flavonoids, alkaloids, tannins, phenolic compounds, steroids, anthraquinones, terpenes, hydrocarbons, volatile oils, glycosides, carbohydrates, proteins and resins. These compounds have already been identified throughout the plant in different concentrations [8, 17, 22, 25-27].

Several secondary metabolites isolated and identified from *T. stans* show activity in several assays. This species contains monoterpene alkaloids, and two of these, tecomine and tecostanine, are apparently responsible for the hypoglycemic activity [28, 29]. Phytochemical investigation of ethanol extract from fruits of *T. stans* has led to the isolation of the monoterpene alkaloids 7-hydroxyskytanthine and 4-hydroxytecomanine, besides 5-hydroxyskytanthine and tecomanine [30].

Constantino et al. [31] isolated the alkaloids tecomine, 5β-hydroxyskytanthine, boschniakine and 5β-hydroxyskytanthine from *T. stans* leaves. Al-Azzawi et al. [32] also isolated from *T. stans* leaves the alkaloids tecomine, dihydrotecomine, boschniakine, 7-hydroxyskytanthine, 5-hydroxyskytanthine and 4-hydroxytecomine.

2.2. Biological Potential of *T. stans*

2.2.1. Allelopathic Effect

Agricultural defenses are widely used by farmers all over the world for the purpose of obtaining great productivity in farming, avoiding pests, weeds and diseases. Due to the high toxicity of herbicides and pesticides, they have become a major hazard to human and animal health and the environment. Therefore, organic agriculture is increasing, and for this it is necessary to use natural herbicides with allelopathic effects, able to safely eliminate planting barriers [33].

According to Carmo et al. [34], allelopathy is defined as the influence of allelochemical compounds that, when released into the environment, affect the development of biological systems. These plant allelochemicals are of extreme importance because they act in ecosystems, influencing plant succession. Several substances are already known for their allelopathic potential, such as polyphenols, flavonoids, tannins and alkaloids, and many have already been characterized in *T. stans*. Cipriani et al. [10] evaluated the allelopathic effect of aqueous extract from the leaves of *T. stans* on the growth and germination of *Lactuca sativa* (cv. Grand Rapids) seeds, and reported 98% inhibition of growth compared to the control group. In another study, Bhat [35] observed the allelopathic potential of aqueous extract from mature leaves of *T. stans* on the seeds of *Vigna radiata* (L.). It caused a remarkable reduction in the growth of the seeds (hypocotyls and radicles); this could be due to allelopathic compounds that influence the development of the roots, besides interfering in the cellular division and metabolization of enzymes of the target species [36].

2.2.2. Antimicrobial Activity

The antimicrobial activity of extracts from the *T. stans* species was evaluated using the qualitative disk-diffusion method and the quantitative minimum inhibitory concentration (MIC) and minimum bactericidal concentration (MBC) methods, verifying the inhibition of various strains of microorganisms: *Staphylococcus aureus*, *Escherichia coli*, *Bacillus subtilis*, *Micrococcus luteus*, *Sarcina lutea*, *Serratia marcescens*, *Salmonella typhi*, *Proteus vulgaris*, *Proteus mirabilis*, *Klebsiella pneumoniae*, *Penicillium* sp., *Aspergillus flavus*, *Cladosporium* sp., *Botrytis* sp. and *Pseudomonas aeruginosa*.

A study carried out by Salem et al. [37] in Mexico verified the antibacterial potential of methanol extract from leaves and branches of *T. stans*, and ethyl acetate (EtOAc), chloroform ($CHCl_3$), n-butanol (n-BuOH) and aqueous (Aq) fractions. Analyses were made by measuring the diameter of growth inhibition zones and MIC values against human pathogenic bacteria (*B. subtilis*, *S. aureus*, *E. coli*, *S. marcescens*, *P.*

aeruginosa, *P. vulgaris*, *S. typhi*, *M. luteus* and *S. lutea*). All microorganisms tested were susceptible to samples from leaves and branches of *T. stans*, to different degrees. $CHCl_3$ and EtOAc fractions of methanol extract from leaves and branches of *T. stans* showed the strongest antibacterial activity, with MIC < 100 µg/mL for *B. subtilis*, and inhibition halos of 20.33 ± 1.52 and 21 ± 1.00 mm at 2000 µg/mL, respectively, for $CHCl_3$ fractions of leaves and branches [37].

In addition, methanol extracts from the leaves and branches inhibited the growth of *B. subtilis* at the concentration of 500 µg/mL. MICs < 100 µg/mL were presented by the $CHCl_3$ fraction of leaves and the methanol extract of branches against *S. aureus*; the n-BuOH fraction of branches against *E. coli* (23.33 ± 1.45 mm); the methanol extract and $CHCl_3$ fraction of shoots and methanol extract of leaves against *S. marcescens*; and the $CHCl_3$ fraction of leaves against *P. aeruginosa*. The $CHCl_3$ and EtOAc fractions of branches and leaves of *T. stans* exhibited the best MIC against *P. vulgaris*: 250 µg/mL for branch fractions and 500 µg/mL for leaf fractions. The methanol extract and $CHCl_3$ fraction inhibited the growth of *S. typhi*, with MICs of 250 and 500 µg/mL, respectively. Methanol extract of leaves inhibited growth of *M. luteus* (MIC = 250 µg/mL), and the EtOAc fraction of branches had activity against *S. lutea* (MIC = 250 µg/mL) [37].

Al-Azzawi et al. [32] evaluated the antibacterial activity of *T. stans* ethanolic and aqueous extracts and tecomine isolated from the leaves. The tecomine presented antibacterial action against *E. coli* (MIC/MBC = 50 µg/mL) and the ethanolic extract against *B. subtilis* (MIC/MBC = 30 µg/mL). However, the aqueous extract was not effective against the microorganisms tested in this study.

Rajamurugan et al. [26] also evaluated the antimicrobial potential of the ethanolic extract of *T. stans* flowers using the disk-diffusion method; it exhibited strong antibacterial activity against all the five bacterial strains, with the maximum inhibitory effect recorded for *E. coli* (23 mm), followed by *K. pneumoniae* (21 mm), *B. subtilis* (21 mm), *P. mirabilis* (19 mm) and *S. aureus* (14 mm). Among the fungi tested, the ethanolic extract of *T. stans* was effective only against *Penicillium* sp. (25 mm).

The antifungal activity of *T. stans* was also evaluated in Brazil. In this work, a trypsin inhibitor (TesTI) was purified, characterized and investigated for anti-*Candida* activity. The TesTI was isolated from the saline extract of *T. stans* leaves, and the antifungal activity was evaluated by MIC and minimal fungicidal concentration (MFC). *Candida* cells treated with TesTI were evaluated for intracellular ATP levels and lipid peroxidation. TesTI inhibited the growth of *C. albicans* and *C. krusei*, with an MIC of 100 μg/mL. The MFC was 200 μg/mL for *C. albicans* and *C. krusei*. Time-response curves revealed that TesTI (in the MIC test) was more effective in inhibiting cell replication in *C. albicans*. In the MIC test, TesTI promoted a reduction of ATP levels and lipid peroxidation in *Candida* cells [38].

2.2.3. Antinociceptive and Anti-Inflammatory Activity

The antinociceptive and anti-inflammatory potential of *T. stans* was studied with the alcohol and aqueous extracts from the leaves. Antinociceptive activity was investigated using the hot plate method, contortion induced by acetic acid, and formalin-induced licking. Anti-inflammatory activity was assessed using the carrageenan-induced paw edema method [27]. The alcohol extract (500 mg/kg) presented greater inhibition of inflammation, 76.92% after 24 h. Both the extracts increased the latency time compared to the vehicle, but the alcohol extract showed higher activity after 150 min in the hot plate method (4.63 ± 0.08 s) and inhibited the nociceptive response in both stadiums. The alcohol and aqueous extracts exhibited a significant dose-dependent decrease in the number of abdominal contortions, the former producing 36.5% and 53.5% inhibition of contortion response at the low and high doses, respectively. These results demonstrate that the alcohol extract of *T. stans* leaves has antinociceptive and anti-inflammatory activity, which may be due to the presence of phenolic compounds and flavonoids [27].

2.2.4. Antioxidant and Anticancer Effects

Reactive oxygen species (ROS) are directly associated with neurodegenerative diseases and early aging, besides being involved as a risk factor for cardiovascular diseases and cancer. However, herbal medicines can be used to prevent diseases associated with the activity of ROS for their antioxidant effect [39].

The antioxidant effect of extract and fractions from leaves and branches of *T. stans* was evaluated (EtOAc, $CHCl_3$, n-BuOH and Aq fractions) using the free radical 2,2-diphenyl-1-picrylhydrazyl (DPPH) method. The highest antioxidant activity of 200 μg/mL was observed for the EtOAc fraction from leaves (83.4 ± 0.31%) and branches (82.06 ± 0.54%), followed by those for the $CHCl_3$ fraction from leaves (79.17 ± 0.20%) and methanol extract from leaves (74.16 ± 0.85%) and branches (72.85 ± 0.31%) [37].

Marzouk et al. [40] reported that *T. stans* has good antioxidant activity. Govindappa et al. [9] evaluated the antioxidant activity of the ethanol and methanol extracts of *T. stans* and their fractions. Ethanol and methanol extracts had higher activity at a concentration of 0.1 mg/mL, with 56.88% and 58.92% DPPH inhibition, respectively [9]. A study realized *in vivo* by Ittagi et al. [41] investigated the cardioprotective effect of 70% ethanolic extract from flowers of *T. stans* on myocardial infarction induced by isoproterenol in rat myocardium. The results indicated that pretreatment with 70% ethanolic extract prevents a decrease in antioxidants and delays significantly the elevation of cardiac damage markers [41].

The anticancer activity of the crude ethanolic extract from leaves of *T. stans* on cancer cells (MCF-7) was evaluated using different concentrations by the 3-(4,5-dimethylthiazol-2-yl)-2,5-diphenyltetrazolium (MTT) assay. The extract showed significant antiproliferative activity, with a dose effect observed. The minimal inhibition was demonstrated at a concentration of 7.8 μg/mL (14.6%), and maximum inhibition (95.9%) at 1000 μg/mL. The authors concluded that the plant extract exhibits potential for additional investigation in cancer cells [25].

2.2.5. Antidiabetic Potential

Giovannini et al. [42] carried out a literature review of the use of plants as traditional remedies, and among the species reported, highlighted the use of *T. stans* for the treatment of Diabetes mellitus.

T. stans is widely used for the empirical treatment of Diabetes mellitus, but its antidiabetic mechanisms still need to be explained [43]. Alonso-Castro et al. [43] evaluated the effect of aqueous extract from leaves of *T. stans* (TSE) on glucose uptake into adipose cells. The non-toxic concentrations were evaluated in adipogenesis, and the uptake of 2-NBD glucose (2-NBDG) evaluated in murine insulin-sensitive, human insulin-resistant and subcutaneous human adipocytes. It was noted that the aqueous extract of *T. stans* stimulated the capture of 2-NBDG by insulin-sensitive and insulin-resistant adipocytes in a concentration-dependent manner. In insulin-sensitive cells, the extract at 70 g/mL stimulated the capture of 2-NBDG by 193% (murine) and 115% (human). In adipocytes resistant to insulin, TSE induced absorption of 2-NBDG by 94% (murine) and 70% (human), compared to the incorporation shown by adipocytes sensitive to insulin stimulated with the hormone. On the other hand, TSE exerted only minimal or null proadipogenic effects in human pre-adipocytes. The conclusion from these results is that *T. stans* exerts its antidiabetic effects by stimulating glucose uptake in both human and murine adipocytes, sensitive and insulin-resistant, without significant proadipogenic or anti-adipogenic side effects [43].

Aguilar-Santamaría et al. [44] demonstrated that the aqueous extract from *T. stans* leaves modulates intestinal glucose absorption in rats, suggesting that the main antidiabetic effect of the plant is due to intestinal alpha-glucosidase inhibition by decreasing the postprandial hyperglycemic peak; in addition, sub-chronic administration of *T. stans* reduces triglycerides and cholesterol, without modifying fasting glucose. Another study suggested that the use of *T. stans* could be in part due to its lipase inhibitory activity, allowing adaptation of the administration of this plant before meals, as auxiliary treatment for type 2 Diabetes mellitus [45].

2.2.6. Nephrotoxicity

Raju et al. [46] evaluated the nephroprotective activity of the ethyl acetate fraction of the dried flowers of *T. stans* on gentamicin-induced nephrotoxicity in albino rats. For the acute toxicity study, a single oral dose of 5.000 mg of fraction/kg of body weight was administered to albino rats.

Nephrotoxicity was induced in albino rats by intraperitoneal administration of 80 mg/kg/day gentamicin for 8 days, with concomitant administration of ethyl acetate fraction at 100, 200 and 300 mg/kg/day administered orally [46].

The results showed that there was no toxicity in general terms: behavior change, mortality or alteration in macroscopic appearance of internal organs (LD_{50} (lethal dose) > 5.000 mg/kg). It was observed that the extract significantly protected rat kidneys from histopathological changes induced by gentamicin. Gentamicin-induced glomerular congestion, peritubular and blood vessel congestion, epithelial desquamation, inflammatory cell accumulation and renal cell necrosis were reduced in the groups that received the ethyl acetate fraction from flowers of *T. stans*, along with gentamicin, in a dose-dependent manner. The floral fraction also reduced the serum creatinine, serum uric acid, urea nitrogen in the blood, and serum urea levels induced by gentamicin [46].

Conclusion

The antimicrobial, anti-inflammatory, antidiabetic, antinociceptive, antioxidant and allelopathic potential and weak nephrotoxicity of *T. stans* have been confirmed in different studies with extracts and fractions, evidencing that this species could be a promising source of new drugs with important biological activity.

Acknowledgments

The authors are grateful to the Federal University of São João Del Rei (UFSJ), the National Council for Scientific and Technological Development (CNPq), Coordination of Improvement of Higher Level Personnel (CAPES) and the Foundation for Research Support of the State of Minas Gerais (FAPEMIG) for their fellowships.

References

[1] Fischer, E., Theisen, I., Lohmann, L. G. (2004). Bignoniaceae. In: Kubitzki, K., Kadereit, J. W. (orgs.). *The families and genera of vascular plants.* Berlin: Heidelberg, 7, 9-98.

[2] Judd, W. S., Campbell, C. S., Kellogg, E. A., Stevens, P. F., Donoghue, M. J. (2002). *Plant systematics: A phylogenetic approach.* Sunderland: Sinauer Associets, 2 ed., 565p.

[3] Angiosperm Phylogeny Group. (2003). An update of the angiosperm phylogeny group classification for the orders and families of flowering plants: APG II. *Bot. J. Linn. Soc.*, 141, 399-436.

[4] Fonseca, S. G. C., Braga, R. M. C., Santana, D. P. (2003). Lapachol - chemistry, pharmacology and assay methods. *Braz. J. Pharmacogn.*, 84, 9-16.

[5] Araújo, R. S. (2008). *Bignoniaceae Juss. do Parque Estadual do Itacolomi, Minas Gerais, Brasil: florística, similaridade e distribuição geográfica.* [*Bignoniaceae Juss. of the Itacolomi State Park, Minas Gerais, Brazil: floristic, similarity and geographical distribution*]. Dissertation (Master in Botanic), Viçosa.

[6] Bredow, E. A. (2007). *Atributos do solo em áreas de ocorrência de Tecoma stans (L.) Juss. ex Kunth (Bignoniaceae) no Estado do Paraná.* [*Soil attributes in areas of occurrence of Tecoma stans (L.) Juss. ex Kunth (Bignoniaceae) in the Parana State*]. Dissertation (Master of Soil Sciences), Curitiba.

[7] Karou, D., Savadogo, A., Canini, A., Yameogo, S., Montesano, C., Simpore, J., Colizzi, V., Traore, A. S. (2005). Antibacterial activity of alkaloids from *Sida acuta*. *Afr. J. Biotechnol.*, 4, 1452-1457.

[8] Raju, S., Kavimani, S., Rao, V. U. M., Reddy, K. S. (2011). *Tecoma stans* (L.) Juss. ex Kunth (*Bignoniaceae*): Ethnobotany, Phytochemistry and Pharmacology. *J. Pharm. Biomed. Sci.*, 8, 1-5.

[9] Govindappa, M., Sadananda, T. S., Channabasava, R., Jeevitha, M. K., Pooja, K. S., Raghavendra, V. B. (2011). Antimicrobial, antioxidant activity and phytochemical screening of *Tecoma stans* (L.) Juss. ex Kunth. *J. Phytol.*, 3, 68-76.

[10] Cipriani, F. A., Kaplan, M. A. C., Isaias, R. M. S., Soares, G. L. G. (2014). Evaluation of the Phytotoxicity of *Tecoma stans* (L.) Kunth. *Floresta e Ambiente* [*Forest and environment*], 21, 1-7.

[11] Renó, L. R. (2002). *Anatomia da raiz e fenologia de Tecoma stans (L.) Kunth (Bignoniaceae).* [*Root anatomy and phenology of Tecoma stans (L.) Kunth (Bignoniaceae)*]. Dissertation (Master of Agronomy), Maringá.

[12] Renó, L. R., Moscheta, I. S., Braccini, A. L. (2007). Morphology and anatomy of the fruit and seed of yellow trumpet flower (*Tecoma stans* (L.) Kunth -*Bignoniaceae*). *Rev. Bras. Sementes*, 29, 18-30.

[13] Santos, J. M. A. (2013). *Apifauna visitante de Tecoma stans (Bignoniaceae) em uma área urbana de João Pessoa, Paraíba, Brasil.* [*Visiting Apifauna of Tecoma stans (Bignoniaceae) in an urban area of João Pessoa, Paraiba, Brazil*]. Monography (Bachelor of Biological Sciences), João Pessoa.

[14] Alves, G. R., Peruchi, A., Agostini, K. (2010). Pollination in urban area: the *Jacaranda mimosifolia* D. Don (*Bignoniaceae*) case study. *Bioikos*, 24, 31-41.

[15] Neto, G. G., Guarim, V. L. M. S., Oliveira, C. A. (2012). Composition of the urban vegetation of cities of the Amazon region - Mato Grosso, Brazil. *Amazônia: Ci. & Desenv.*, 7, 137-155.

[16] Kranz, W. M., Passini, T. (1997). Amarelinho: biologia e controle. *Informe da Pesquisa* [Yellowing: biology and control. *Research Report*], 121, 1-19, Instituto Agronômico do Paraná.

[17] Dash, S., Das, C., Sahoo, D. C., Sahoo, A. C. (2011). Phytochemical composition, anti-inflammatory and analgesic activities of *Tecoma stans* Linn. (*Bignoniaceae*). *Nat. Pharm. Technol.*, 1, 5-8.

[18] Ramesh, T., Anusha, V., Kumar, A. R. (2009). Antibacterial activity of methanolic extract of roots of *Tecoma stans*. *Int. J. Chem. Sci.*, 7, 6-8.

[19] Singh, V., Kumar, L., Chakraborthy, G. S., Mazumder, A. (2011). Pharmacological and phytochemical findings of *Tecoma stans.* A review. *J. Adv. Pharm. Healthcare Res.,* 1, 75-81.

[20] Orwa, C., Mutua, A., Kindt, R., Jamnadass, R., Simons, A. (2009). *Agroforestree database: a tree reference and selection guide version 4.0.* World Agroforestry Centre, Kenya. Available from: http:// www.worldagroforestry.org/output/agroforestree-database. Accessed on November, 16, 2018.

[21] Bin Asad, M. H. H., Razi, M. T., Sabih, D., Najamus-Saqib, Q., Nasim, S. J., Murtaza, G., Hussain, I. (2013). Anti-venom potential of Pakistani medicinal plants: inhibition of anticoagulation activity of *Naja naja karachiensis* toxin. *Curr. Sci.*, 105, 1419-1424.

[22] Das, C., Dash, S., Sahoo, D. C., Mohanty, A. (2010). Evaluation of metabolic bark extract of *Tecoma stans* Linn, for wound healing in albino rats. *Int. J. Pharm. Technol.*, 2, 735-742.

[23] Egwaikhide, P. A., Gimba, C. E. (2007). Analysis of the phytochemical content and antimicrobial activity of *Plectranthus glandulosis* whole plant. *Middle East J. Sci. Res.*, 2, 135-138.

[24] Binutu, O. A., Lajubutu, B. A. (1994). Antimicrobial potentials of some plant species of the *Bignoniaceae* family. *Afr. J. Med. Med. Sci.*, 23, 269-273.

[25] Thirumal, M., Kishore, G., Prithika, R., Das, S., Nithya, G. (2012). In vitro anticancer activity of *Tecoma stans* (L.) ethanolic leaf extract on human breast cancer cell line (MCF-7). *Int. J. Pharm. Chem. Biol. Sci.*, 2, 488-493.

[26] Rajamurugan, R., Thirunavukkarasu, C., Sakthivel, V., Sivashanmugam, M., Raghavan, C. M. (2013). Phytochemical screening, antioxidant and antimicrobial activities of ethanolic extract of *Tecoma stans* flowers. *Int. J. Pharm. Bio. Sci.,* 4, 124-130.

[27] Prasanna, V. L., Lakshman, K., Hedge, M. M., Bhat, V. (2013). Antinociceptive and anti-inflammatory activity of *Tecoma stans* leaf extracts. *Indian J. Res. Pharm. Biotechnol.,* 1, 156-160.

[28] Hammouda, Y., Amer, M. S. (1966). Antidiabetic effect of Tecomine and Tecostanine. *J. Pharm. Sci.*, 55, 1452-1454.

[29] Hammouda, Y., Rashid, A. K., Amer, M. S. (1964). Hypoglycaemic properties of Tecomine and Tecostanine. *J. Pharm. Pharmacol.*, 16, 833-834.

[30] Lins, A. P., Felício, J. D. (1993). Monoterpene alkaloids from *Tecoma stans*. *Phytochem.*, 34, 876-878.

[31] Costantino, L., Raimondi, L., Pirisino, R., Brunetti, T., Pessotto, P., Giannessi, F., Lins, A. P., Barlocco, D., Antolini, L., El-Abady, S. A. (2003). Isolation and pharmacological activities of the *Tecoma stans* alkaloids. *Il Farmaco,* 58, 781-785.

[32] Al-Azzawi, A. M., Al-Khateeb, E., Al-Sameraei, K., Al- Juboori, A. G. (2012). Antibacterial activity and the histopathological study of crude extracts and isolated tecomine from *Tecoma stans Bignoniaceae* in Iraq. *Pharmacognosy Res.,* 4, 37-43.

[33] Moraes, R. M., Cerdeira, A. L., Duke, S. O., Dayan, F. E., Cantrell, C., Queiroz, S. C. N. (2016). Pesticidas Naturais Derivados de Plantas: Descoberta e Usos. [Natural Pesticides Derived from Plants: Discovery and Uses]. In: Halfeld-Vieira, B. A., Marinho-Prado, J. S., Nechet, K. L., Morandi, M. A. B., Bettiol, W. (Ed.). *Defensivos Agrícolas Naturais*. Embrapa, 1, 505-541.

[34] Carmo, F. M. S., Borges, E. E. L., Takaki, M. (2007). Allelopathy of Brazilian sassafras (*Ocotea odorifera* (Vell.) Rohwer) aqueous extracts. *Acta Bot. Bras*., 21, 697-705.

[35] Bhat, M. A., Yogamoorthi, A. (2017). Evaluation of allelopathic effect of aqueous leaf extract of *Tecoma stans* (L.) on seed germination and biochemical changes in *Vigna radiata* (L.). *Int. J. Curr. Res*., 9, 51014-51021.

[36] Pawlowski, A., Kaltchuk-Santos, E., Zini, C. A., Caramão, E. B., Soares, G. L. G. (2012). Essential oils of *Schinus terebinthifolius* and *S. molle* (Anacardiaceae): Mitodepressive and aneugenic inducers in onion and lettuce root meristems. *S. Afr. J. Bot*., 80, 96-103.

[37] Salem, M. Z. M., Gohar, Y. M., Diaz, L. M. C., El-Shanhorey, N. A., Salem, A. Z. M. (2013). Antioxidant and antibacterial activities of leaves and branches extracts of *Tecoma stans* (L.) Juss. ex Kunth against nine species of pathogenic bacteria. *Afr. J. Microbiol. Res*., 7, 418-426.

[38] Patriota, L. L. S., Procópio, T. F., Souza, M. F. D., Oliveira, A. P. S., Carvalho, L. V. N., Pitta, M. G. R., Rego, M. J. B. M., Paiva, P. M. G., Pontual, E. V., Napoleão, T. H. (2016). A Trypsin Inhibitor from *Tecoma stans* Leaves Inhibits Growth and Promotes ATP Depletion and Lipid Peroxidation in *Candida albicans* and *Candida krusei*. *Front. Microbiol.*, 7, 1-10.

[39] Buchli, L. (2002). Radicais livres e antioxidantes. [Free radicals and antioxidants]. *Cosmetics & Toiletries*, 14, 54-57.

[40] Marzouk, M., Gamal-Eldeen, A., Mohamed, M., El-Sayeed, M. (2006). Antiproliferative and antioxidant constituents from *Tecoma stans*. *Z. Naturforsch C.*, 61, 783-791.

[41] Ittagi, S., Merugumolu, V. K., Siddamsetty, R. S. (2014) Cardioprotective effect of hydroalcoholic extract of *Tecoma stans* flowers against isoproterenol induced myocardial infarction in rats. *Asian Pac. J. Trop. Dis.*, 4, 378-384.

[42] Giovannini, P., Howes, M. J. R., Edwards, S. E. (2016). Medicinal plants used in the traditional management of diabetes and its sequelae in Central America: A review. *J. Ethnopharmacol.*, 184, 58-71.

[43] Alonso-Castro, A. J., Zapata-Bustos, R., Romo-Yañez, J., Camarillo-Ledesma P., Gómez-Sánchez, M., Salazar-Olivo, L. A. (2010). The antidiabetic plants *Tecoma stans* (L.) Juss. ex Kunth (*Bignoniaceae*) and *Teucrium cubense* Jacq (*Lamiaceae*) induce the incorporation of glucose in insulin-sensitive and insulin-resistant murine and human adipocytes. *J. Ethnopharmacol.*, 127, 1-6.

[44] Aguilar-Santamaría, L., Ramírez, G., Nicasio, P., Alegría-Reyes, C., Herrera-Arellano, A. Antidiabetic activities of *Tecoma stans* (L.) Juss. ex Kunth. (2009). *J. Ethnopharmacol.*, 124, 284-288.

[45] Ramirez, G., Zamilpa, A., Zavala, M., Perez, J., Morales D., Tortoriello, J. (2016). Chrysoeriol and other polyphenols from *Tecoma stans* with lipase inhibitory activity. *J. Ethnopharmacol.*, 185, 1-8.

[46] Raju, S., Kavimani, S., Rao, V. U. M., Reddy, K. S., Kumar, G. V. (2011). Floral extract of *Tecoma stans*: A potent inhibitor of gentamicin-induced nephrotoxicity in vivo. *Asian Pac. J. Trop. Med.*, 4, 680-685.

In: Phytochemicals
Editor: Iman Ryan
ISBN: 978-1-53615-478-8

Chapter 10

BIOLOGICAL POTENTIAL OF *PYROSTEGIA VENUSTA* (KER GAWL.) MIERS (*BIGNONIACEAE*): A REVIEW

Mairon César Coimbra, Ana Hortência Fonsêca Castro and Luciana Alves Rodrigues dos Santos Lima*

Campus Centro-Oeste Dona Lindu, Federal University of São João Del Rei, Divinópolis, Minas Gerais, Brazil

ABSTRACT

Many plant species produce bioactive compounds, which can play an important role in human health. *Pyrostegia venusta* (Ker Gawl.) Miers (synonyms: *Pyrostegia ignea* and *Bignonia venusta*) belongs to the *Bignoniaceae* family and is a wood-climbing vine dispersed in tropical regions. *Bignoniaceae* Juss. is represented by about 82 genera and 800 tree species, shrubs and creepers, with pantropical distribution and pronounced occurrence in the neotropics. The family is subdivided into eight tribes which are differentiated by plant habit and fruit morphology. Brazil has some endemic taxa and is considered the center of diversity of

* Corresponding Author's E-mail: luarsantos@ufsj.edu.br.

the family, with 32 genera and 350 species occurring in several types of environment from the Brazilian Cerrado to perennial wet forests. *P. venusta* is a seasonal species easily recognized by the abundant production of orange flowers.

In folk medicine, *P. venusta* is used to treat coughs, bronchitis, infections, erysipelas, jaundice, vitiligo and diarrhea. Ethnopharmacological surveys consistently describe its potential as a medicinal plant, and much pharmacological activity is related to phenolic compounds, mainly flavonoids, such as hesperidin, homoeriodictyol, rutin, quercetin, catechin and acacetin-7-*O*-glycoside found in leaves. Other bioactive compounds are found in flowers and roots, such as carotenoids, β-amyrin, β-sitosterol, *myo*-inositol, oleanolic acid, hexadecanoic acid, linoleic acid, oleic acid and allantoin. Recent studies showed an absence of genotoxic activity in the ethanolic extracts of *P. venusta* flowers, and good ability of methanolic extracts of leaves, flowers and roots to scavenge free radicals. Other biological activity has been reported, such as the improvement of flu symptoms, stimulation of melanogenesis, wound healing and anti-inflammatory, antinociceptive and antimicrobial activity.

P. venusta is an interesting source of chemiodiversity, and the compounds obtained from this plant can be used as functional foods, drugs or precursors for the semi-synthesis of drugs. So, this plant represents an important source of bioactive compounds of pharmaceutical/medicinal interest, with recognized biological potential.

Keywords: biopharmaceuticals, biological activity, medicinal plant, *Pyrostegia venusta*, secondary metabolites

1. THE *BIGNONIACEAE* FAMILY AND THE *PYROSTEGIA* GENUS

Bignoniaceae is a family of woody plants with 82 genera and approximately 860 species [1]; about 800 of the species are trees, while the rest are distributed among lianas, shrubs and herbs. In the most recent classifications of angiosperms, *Bignoniaceae* is inserted in the order Lamiales [2, 3]. The family is pantropical and occurs abundantly in tropical regions of South America, with only a few genera reaching temperate climates or high mountains. Specimens of *Bignoniaceae* are generally recognized as being woody, and they have opposite leaves, tubular and zygomorphic flowers, with four didynamous stamens plus a staminode and bi-valved dry capsular fruits [4-7]. Eight tribes were recognized based on habit, fruit dehiscence and geographical distribution: Bignonieae, Coleeae, Crescentieae, Eccremocarpeae, Oroxyleae, Tecomeae, Tourrettieae and Schlegelieae [4, 6]. Recently, molecular phylogenetic studies have indicated that some tribes were not monophyletic as traditionally circumscribed, suggesting that nine major clades should be recognized in their place [7]. Of these nine clades, two correspond to tribes previously recognized under the Gentry system [4], that is, Bignonieae and Oroxyleae. Two other tribes previously recognized under the Gentry system (1980), Crescentieae and Coleeae, also emerged as monophyletic but nested within more inclusive clades; Crescentieae emerged within the clade of the Neotropical Tabebuia alliance [8], and Coleeae emerged within the Paleotropical clade [7]. The Tecomeae tribe appeared dispersed in six different clades: Catalpeae, the Tabebuia alliance, the Paleotropical clade, Delostoma, Tecomeae and Jacarandeae. Although these clades are well supported by molecular characteristics, additional morphological and anatomical features, such as the anatomy of wood, are still desirable to characterize them [9].

Pyrostegia C. Presl is a genus of four species of the Bignonieae tribe. All species are lianas that have leaves composed of two leaflets and a terminal ridge, campanulate chalices, narrow corollas with valved lobes at the base, four exposed stamens, linear compressed capsules that develop parallel to the septum, and fine and bialate seeds. Three of the species are apparently very similar, presenting orange and narrow tubular-infundibular corollas. The fourth species has white salveriform corollas. All four species are native to South America. Lectotypes are called *Bignonia ignea* Vell., *Bignonia tecomiflora* Rusby, *Pyrostegia cinerea* Bureau ex K. Schum., *Pyrostegia dichotoma* Miers ex K. Schum., *Pyrostegia venusta* var. villosa Hassl. and *Tynanthus igneus* Barb. Rodr. [10].

2. The *Pyrostegia venusta* Species

Pyrostegia venusta (Ker Gawl.) Miers (Figure 1), synonyms *Pyrostegia ignea* and *Bignonia venusta*, is popularly known in Brazil as "cipó-de-São-João," "erva-de-São-João" and "flor-de-São-João." These names are allusive to its constant use in the decoration of the Catholic festivities of São João, from north to south of the whole country [11]. In other locations it can be commonly named "flame vine" or "orange trumpet creeper" [12, 13]. *P. venusta* is a perennial, climbing, woody plant, measuring from 2 to 4 m long, with seed reproduction; it has opposite leaves, composed of two leaflets and a trifid tendril, possibly with three leaflets, 3 cm long and 2 cm wide. Its flowers are axillary in the form of panicles, with tubular corollas, of orange color and are arranged in bunches with 15 to 20 flowers on average [12].

P. venusta is a vigorous, fast-growing woody vine that blooms in winter and spring with showy orange blossoms, which have even been included, as *Pyrostegia ignea*, in a list of vines with the most beautiful blooms in the world and classified as popular in the tropics [13, 14]. *P. venusta* can be found in tropical and subtropical areas; it is native to southern and southeastern Brazil, northern Argentina and Paraguay. Flame vine is characterized by adapting well in environments of acidic and alkaline soils ranging from hygrophyte to mesophyte, being easily found in forest edges, fields, coasts and roadsides [13, 15]. In some regions of its occurrence, one can find *P. venusta* being used as an ornamental plant covering fences and walls [13, 15, 16].

In folk medicine, *P. venusta* is widely used by decoction or infusion of branches with leaves and/or flowers to treat coughs, bronchitis, vitiligo and diarrhea [17-20], as well as in the treatment of uterine and female genital tract infections, erysipelas and jaundice (hyperbilirubinemia and deposition of bile pigments) in newborn babies [21].

Natural products, especially those of plant origin, have been the source of most active drug principles [22]. *P. venusta* has been the subject of recurrent phytochemical investigations. In the flowers were isolated several compounds such as β-amyrin, β-sitosterol, *n*-hentriacontane (*n*-

$C_{31}H_{64}$), stigmasteryl tosylate, diazoprogesterone, arabinopyranose, propanoic acid, pentamethyldisilanyl ester, acetophenone, 3-hexenedioic acid, methyl ester 9-octadecenoic acid, terpenes, sterols, flavonoids, condensed tannins, saponins, coumarins, anthraquinones, carotenoids, acacetin-7-*O*-*β*-glucopyranoside, *myo*-inositol, hexadecanoic acid, linoleic acid, palmitic acid, quercetin-3-*O*-L-aminopyranosyl-(1→6)-*β*-D-galactopyranoside, oleic acid, allantoin and phenylpropanoid glycosides (isoverbascoside and verbascoside) [23-33].

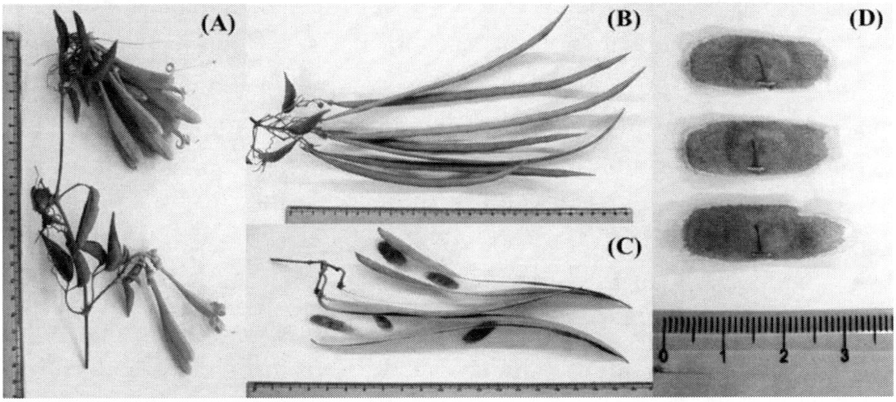

Figure 1. *Pyrostegia venusta* (Ker Gawl.) Miers. (A) fertile branches, (B) green fruit, (C) mature fruit evidencing its characteristic dehiscence, (D) seeds.

For the leaves, sterols, coumarins, flavones, triterpenes, saponins, phenols, flavonoids, allantoin, rutin, syringyl groups, quercetin, catechin and biochanin have been reported [28, 34-37]. Choline chloride has been found in the bark of the stem [38]. In the roots are found saponins, tannins, terpenes, alkaloids, allantoin, *β*-sitosterol, 3-*β*-*O*-*β*-D-glucopyranosylsitosterol and the flavanone hesperidin [17, 29]. For the seeds, there are reports of fatty acids [39].

Compounds such as steroids, phenols, flavonoids and gallic, benzoic, ferulic and *p*-coumaric acids have also been found in biotechnological products (calli) derived from the leaves of *P. venusta* [37, 40, 41].

Although there are reports of the presence of iridoids in several species of the *Bignoniaceae* family, a survey by von Poser et al. [42] did not find the presence of this class of substances in *P. venusta*.

3. *PYROSTEGIA VENUSTA*: BIOTECHNOLOGICAL APPROACH

Plants can be used to produce a wide variety of complex molecules with industrial potential or pharmaceutical use. However, there is a difference in the production of these compounds at laboratory and industrial scale. Thus, if the focus of the production is molecules that have high commercial potential aggregate, aspects such as quality, purity and crop yield should be considered, rather than simply obtaining the target molecule [43].

The biotechnological potential of a medicinal species is related to the ability of its cells to produce secondary metabolites under the appropriate physical or chemical stimuli. In this context, *in vitro* plant cell culture emerges as an effective biotechnological alternative for the production of substances with therapeutic potential [44, 45].

The success of the cell culture depends on the asepsis conditions in which the seeds and explants are cultivated. Observation of these factors is necessary in order to resolve or cancel the involvement of fungi or bacteria in the biosynthesis of the target metabolite. Contamination by microorganisms is considered to be one of the most important factors causing a decrease in the productivity of *in vitro* cultures, because these agents compete directly for the availability of water and nutrients present in the culture medium [46]. With this in mind, Coimbra and Castro [47] established a simple, fast, economical and efficient protocol for disinfestation of *P. venusta* leaf explants in order to maintain their viability for subsequent callus induction studies.

P. venusta can be cultured in a conventional manner [11, 20, 48, 49]. However, *in vitro* propagation is advantageous because it provides aseptic conditions and more complete nutrient supplementation for faster and more efficient germination and for seedling growth [50]. In addition, pre-treatments can be applied to seeds and stem segments in order to optimize plant growth and development. The *in vitro* propagation method is independent of the climate and seasonal variations, capable of producing the required amount of seedlings at a given time, and does not require extensive fieldwork. Braga et al. [41] obtained *in vitro* germination rates for *P. venusta* greater than 80% for different types and concentrations of culture medium. The seedlings obtained in this experiment presented total phenolic and flavonoid values in the order of 8.50 μg of tannic acid equivalents and 3.80 μg of rutin equivalents per milligram of dry matter, respectively.

Among the *in vitro* culture techniques, callus culture is a viable alternative for the preservation of plant species and for obtaining bioactive metabolites [41]. Calli are tissues that have partial differentiation, consisting of a mass of irregular cells that multiply in response to chemical stimuli or physical injuries and have the capacity to differentiate into tissues and organs [51]. Coimbra et al. [37], Braga et al. [41] and Loredo-Carrillo et al. [40] induced calli from leaf explants of *P. venusta* and obtained phenolic compounds and phytosterols. Loredo-Carrillo et al. [40] showed that water stress increased the sterol content 4.6 times, while osmotic stress induced a 3.8- and 2.2-fold increase in the concentration of sterols and flavonoids, respectively, in relation to the leaves of the plant donor of explants.

Coimbra et al. [37] showed that the presence or absence of light, as well as the type and concentration of the growth regulator used for callus induction, can influence the production and accumulation of secondary metabolites in *P. venusta* calli. This work also highlights that the combined use of auxin and cytokinin, in the absence of light, produces friable callus with a higher phenol and flavonoid content than the initial explant. The presence of caffeic, gallic and benzoic acid derivatives in hydrometanic extracts of *P. venusta* calli has also been reported.

Recently, numerous strategies have been developed to increase plant biomass and the production of secondary metabolites. According to

Murthy et al. [44], these events occur in two stages: control of the parameters related to the growth and multiplication of cells/organs, and control of metabolite biosynthesis. Variables such as the type and appropriate composition of the culture medium, concentrations of carbon sources (carbohydrates), nitrogen, phosphate and plant growth regulators, and physical factors such as luminosity, pH, agitation and aeration can be controlled in the first phase of the culture process. In the second stage, it is possible to use elicitors, precursors, restoration of nutrients and strategies of permeabilization and immobilization of cell cultures so that biosynthesis of the target metabolite is increased [44, 45].

4. BIOLOGICAL POTENTIAL OF *PYROSTEGIA VENUSTA*

4.1. Antimicrobial Activity

P. venusta has shown antibacterial activity against several microorganisms of clinical importance. Hydroethanolic and aqueous extracts from leaves of 20 climbing species of the *Bignoniaceae* were evaluated *in vitro* by Torres et al. [52] against Gram-positive and Gram-negative strains. In this study, hydroethanolic extracts of *P. venusta* leaves showed high antimicrobial activity against *Staphylococcus aureus* (ATCC 29213), *Staphylococcus aureus* (ATCC 25923), *Staphylococcus epidermidis* (ATCC 12228), *Enterococcus faecalis* (ATCC 29212) and clinical isolates of methicillin-sensitive (*Staphylococcus aureus*, *Escherichia coli* and *Pseudomonas aeruginosa*) and antibiotic-resistant species (*Proteus mirabilis* and *Morganella morganii*). However, Fernandes et al. [53] did not find activity against *Bacillus cereus*, *Bacillus stearothermophilus*, *Bacillus subtilis*, *Enterobacter aerogenes*, *Escherichia coli*, *Klebsiella pneumoniae*, *Micrococcus luteus*, *Proteus mirabilis*, *Pseudomonas aeruginosa*, *Salmonella typhimurium*, *Staphylococcus aureus*, *Staphylococcus epidermidis*, *Streptococcus pyogenes*, *Streptococcus salivarius*, *Candida albicans*, *Cryptococcus neoformans* and *Saccharomyces cerevisiae*.

Silva et al. [54] also evaluated the antibacterial activity of organic extracts (hexane, ethyl acetate and methanol) of leaves by MIC (minimum inhibitory concentration). The hexane extract showed moderate inhibition against *Staphylococcus aureus* (MIC = 0.9 mg mL^{-1}) and strong inhibition against *Enterococcus hirae* (MIC = 0.5 mg mL^{-1}). The ethyl acetate extract presented strong activity against *Candida albicans* (MIC = 0.3 mg mL^{-1}), while the methanolic extract was not active for the microorganisms tested. However, Bouzada et al. [28] reported antibacterial activity of the methanolic extract from *P. venusta* leaves and flowers through the agar diffusion assay and showed activity against several pathogenic bacteria such as *Staphylococcus aureus*, *Pseudomonas aeruginosa*, *Salmonella typhimurium*, *Shigella sonnei*, *Klebsiella pneumoniae*, *Escherichia coli* and *Bacillus cereus*.

According to De Souza et al. [55], crude extract, ethyl acetate and *n*-butanol fractions of *P. venusta* flowers exhibit antibacterial and antifungal activity, and the ethyl acetate fraction shows the greatest inhibition of *Streptococcus mutans* and *Candida albicans*. The results of this work suggest that extracts of the flowers can be used as a preliminary source of compounds for prophylaxis or treatment of caries or periodontal disease. An *n*-butanolic fraction, semi-purified from the hydroethanolic extract of *P. venusta* flowers, showed activity similar to amphotericin B and antifungal effects, especially against *Candida* strains (*C. albicans*, *C. krusei*, *C. tropicalis*, *C. parapsilosis* and *C. guilliermondii*) [32]. However, Roy et al. [56] reported the moderate activity of methanolic extracts against *Bacillus subtilis*, *Staphylococcus epidermidis*, *Staphylococcus pyogenes*, *Escherichia coli*, *Micrococcus luteus*, *Enterobacter aerogenes*, *Salmonella typhi*, *Pseudomonas aeruginosa*, *Aspergillus niger* and *Candida tropicana*.

Ethanolic extracts from *P. venusta* stems and leaves were evaluated *in vitro* against human herpesvirus type 1 (HSV-1), vaccinia virus (VACV) and murine encephalomyocarditis virus (EMCV) by 3-(4,5-dimethylthiazol-2-yl)-2,5-diphenyltetrazolium but no activity was observed [57].

4.2. Antioxidant Activity

Some studies suggest that *P. venusta* is a natural source of antioxidants. Methanolic extracts and ethyl acetate fractions exhibit high free radical scavenging ability using DPPH (2,2-diphenyl-1-picryl-hydrazyl), 2,2'-azinobis-3-ethylbenzothiazoline-6-sulfonic acid (ABTS) and Trolox as standards [36, 54].

Flowers and roots of *P. venusta* have significant amounts of phytochemicals such as *myo*-inositol, stigmasteryl tosylate, phthalic acid, hexadecanoic acid, linoleic acid, palmitic acid and oleic acid with antioxidant properties that can be used as inhibitors or free radical scavengers. The antioxidant activity of both extracts is close to those of hydroxytoluene butoxide (BHT), catechin and quercetin. A significant inhibitory effect of flower and root extracts on ABTS free radicals has also been detected. The antioxidant activity of flower extracts (95%) and roots (94%) on DPPH radicals is comparable to that of ascorbic acid (98.9%) and BHT (97.6%) [29]. The role of free radicals in wound pathogenesis is well known. Roy et al. [56] reported that *P. venusta* flowers have potent antioxidant activity, inhibiting lipid peroxidation, reducing glutathione levels and increasing superoxide dismutase (SOD) and catalase (CAT) activity.

The crude extract and *n*-butanolic fraction of *P. venusta* flowers, obtained by Pereira et al. [32], have also been shown to contain promising antioxidant agents such as quercetin-3-*O*-L-aminopyranosyl-(1→6)-*β*-D-galactopyranoside and two glycosylated phenylpropanoids (verbascoside and isoverbascoside), which are able to reduce DPPH as well as rutin. The activity is more evident for the *n*-butanolic fraction, with a 93% reduction of DPPH at concentrations up to 125 μg mL^{-1}.

4.3. Anthelmintic Activity

Nisha et al. [13] evaluated the anthelmintic properties of chloroform and methanolic extracts of *P. venusta* leaves and stems using *Pheretima posthuma* as the experimental model and piperazine citrate as standard. Among the various concentrations of the chloroform extract tested, the concentration of 12.5 mg mL^{-1} showed efficient anthelmintic activity with a paralysis time of 23 min and death time of 44 min. The methanolic extract, at a concentration of 12.5 mg mL^{-1}, showed significant results, with a paralysis time of 34 min and death time of 78 min. This work revealed that the chloroform extract is more effective against *Pheretima posthuma* than the methanolic extract. However, both extracts were less potent when compared to the piperazine citrate used as standard.

4.4. Anti-Inflammatory and Antinociceptive Activity

Veloso et al. [30] evaluated the anti-inflammatory and analgesic effects of hydroethanolic extract of *P. venusta* flowers on carrageenan-induced paw edema, lipopolysaccharide-induced peritonitis, acetic acid-induced writhing, and formalin-induced foot-lick tests in male Swiss mice. The extract showed an anti-inflammatory effect at concentrations between 30 and 300 mg kg^{-1}. It also reduced paw edema and inhibited the recruitment of leukocytes to the peritoneal cavity, as well as showing antinociceptive activity in the tests of contortion induced by acetic acid and formalin.

Veloso et al. [58] reported that the same extract was used to verify the effects of supplementation for the treatment of inflammatory and metabolic dysfunction induced by a diet rich in refined carbohydrates. The extract decreased adiposity and adipocyte area, improved glucose intolerance, reduced serum levels of triacylglycerols and systemic inflammatory cells and also reduced some levels of inflammatory mediators in adipose tissue and liver. The results showed that hydroethanolic extracts of *P. venusta* flowers have beneficial effects and can be used to treat the inflammatory and metabolic dysfunction induced by a diet rich in refined carbohydrates and that are associated with negative modulation of the inflammatory process at systemic and local levels.

The anti-inflammatory activity of hydroethanolic extracts of *P. venusta* leaves on inflammation induced by Croton oil was verified by Moreira et al. [59] in animal models. Involvement in cell infiltration and cytokine levels was evaluated in C56BL/6 mice. *P. venusta* extract increased tissue levels of TNF-α (68.2 ± 11.6% and 99.2 ± 12.1%, respectively) and cell infiltration (baseline levels and 94.3 ± 9.17%, respectively) both by gavage and topical administration.

4.5. Cold and Flu Treatment

P. venusta flower extracts were tested for their effects during oral administration on lipopolysaccharide-induced disease behavior in mice [60]. To evaluate the effects of extract administration, the mice were submitted to the forced swimming test (FST) and the open field test. Pretreatment with the *P. venusta* extract attenuated the behavioral changes induced and observed in the tests, presenting results similar to that of dexamethasone, a steroidal drug that inhibits immune and inflammatory responses, including the production of cytokines. The authors further pointed out that extracts from this plant may be useful in treating disorders that induce unhealthy behavior, such as colds and flu.

4.6. Allelopathic Activity

Allelopathic study investigates the positive and negative effects that secondary metabolites of plants, microorganisms or fungi exert on the development of neighboring individuals. Da Silva et al. [61] reported the effects of the hydroethanolic extract of *P. venusta* flowers on germination and root system development of *Brassica chinensis* L. var. *parachinensis* (Bailey) Sinskaja. Application of different concentrations of the extract elicited changes in the germinative parameters and in the root length of the test plants, indicating the existence of allelopathic potential. The allelopathic potential of *P. venusta* was also studied in *Cucumis sativus*; changes in the main root length and the number of secondary roots due to the hexane, ethyl acetate and methanolic extracts were observed, whereas only the ethyl acetate extract did not affect the length of the hypocotyl [54]. Daneluzzi et al. [33] also reported changes in the germination index and the root length of *Lactuca sativa* (Asteraceae) in the presence of different concentrations of hydroethanolic extracts of *P. venusta* flowers.

4.7. Wound Healing Activity

Roy et al. [56] investigated the effect of methanolic extracts from *P. venusta* flowers in the wound healing process, coupled with their effect on proinflammatory and anti-inflammatory cytokines, through incision and wound repair in Wistar rats. Wound healing was evidenced by the contraction and increase in tensile strength. The content of hydroxyproline and hexosamine was also correlated with the observed healing pattern. During the initial wound healing phase, the levels of proinflammatory cytokines such as tumor necrosis factor-alpha (TNF-α) and interleukin-6 (IL-6) were upregulated by treatment with *P. venusta* extracts.

4.8. Melanogenic Activity

Moreira et al. [31] studied the melanogenic activity of hydroethanolic extract of *P. venusta* leaves and flowers on murine melanoma B16F10 cells. The extracts of leaves (0.1, 0.3, 1 and 3 g mL^{-1}) and flowers (0.03 and 0.1 g mL^{-1}) increased the melanin content in a concentration-dependent manner after 4 days of incubation of melanoma cells. Leaf extract promoted an increase in melanogenesis with a maximum effect of 33.3 ± 3%, while flower extract increased melanogenesis by 23.4 ± 3%. Both extracts stimulated melanogenesis in B16F10 cells at very low concentrations. These findings corroborate the medicinal use of *P. venusta* in the treatment of hypopigmentation diseases such as vitiligo [62]. Moreira et al. [59] reported the hyperpigmentation capacity of the hydroethanolic extract of *P. venusta* leaves in animal models of vitiligo. Involvement in mouse pigmentation was evaluated in the monobenzoneinduced vitiligo model in C56BL/6 mice. The extracts increased the level of epidermal melanin and decreased dermal depigmentation by gavage (300 mg kg^{-1}) or topical treatment (10%). However, only topical treatment altered the specific marker for melanin in the hair follicles.

4.9. Antitumor Activity

Some studies have shown the preventive and therapeutic antitumor activity of *P. venusta* extracts. Silva et al. [63] reported the antitumor action of hydroethanolic extracts from *P. venusta* flowers evaluated by inhibition of *Agrobacterium tumefaciens*-induced tumors in potato tissue disks.

Figueiredo et al. [64] showed the antitumor activity of heptanic partitions of hydroethanolic extracts from *P. venusta* flowers on murine melanoma cells (B16F10-Nex2) and *in vivo* models. The authors observed that there was: a) induction of apoptosis in melanoma cells by rupturing the mitochondrial membrane potential; b) induction of reactive oxygen species; c) late apoptosis evidenced by plasma membrane blebbing; d) cell shrinkage, chromatin condensation and DNA fragmentation; e) phosphatidylserine exposure on the cell surface; and f) activation of caspases. This heptanic partition also protected the *in vivo* models against syngeneic subcutaneous melanoma and was also able to induce cell cycle arrest in the G2/M phases in tumor cells.

4.10. Cytotoxic Activity

Magalhães et al. [65] evaluated the genotoxic effect of hydroethanolic extracts from *P. venusta* flowers in mice using the micronucleus and chromosome aberration tests. The frequency of micronucleated polychromatic erythrocytes of the experimental groups was significantly inferior when compared to the controls. The chromosome aberration had no significant difference when compared to the negative control but was statistically lower than that of the positive control. So, extracts of *P. venusta* did not present genotoxic activity. This result was confirmed also for hydroethanolic extracts from leaves by the micronucleus test in erythroblasts of Swiss albino mice. After 24 and 48 h of the treatment by gavage (1,000-2,000 mg kg^{-1}), the absence of potentially clastogenic and/or aneugenic effects was verified, independent of the sex, time and dosage tested [55]. Bouzada et al. [28] and Silva et al. [63] reported that hydroethanolic and methanolic extracts from *P. venusta* flowers and leaves do not show significant cytotoxicity against *Artemia salina*. According to Daneluzzi et al. [33], the hydroethanolic extracts from flowers also do not have toxic action on the mitotic index in *Lactuca sativa* seeds. The absence of cytotoxicity was observed also by Brandão et al. [52] on Vero cells exposed to different concentrations of the ethanolic extract from leaves and stems for 48 and 72 h after incubation. According Altoé et al. [36], ethanolic extracts from *P. venusta* leaves show lower cytotoxicity against RAW 264.7 macrophages than four commercial anti-inflammatory drugs (acetylsalicylic acid, indomethacin, betamethasone and piroxicam).

Non-pathogenic trypanosomatids such as *Herpetomonas samuelpessoai* provide an important model for the study of basic biological processes carried out by a eukaryotic cell [66]; the species was used by Fernandes et al. as a model for the study of the effects of *P. venusta* extract on cell growth and differentiation. In the presence of hydroethanolic extract, the number of trypanosomatids was not reduced and even increased when the concentration of the extract was increased, showing that the extract does not inhibit the growth of *H. samuelpessoai* at the concentrations tested [54]. This result indicates that hydroethanolic extract of *P. venusta* leaves is safe, since it does not present carcinogenic/ mutagenic potential.

4.11. Angiotensin-Converting Enzyme Inhibitory Activity

Angiotensin-converting enzyme (ACE) plays a key role in the homeostatic mechanism of mammals, being involved in the regulation and control of arterial pressure. ACE inhibitors have Food and Drug Administration (FDA)-approved indications for treatment of high blood pressure, heart failure, left ventricular dysfunction and diabetic nephropathy [67, 68]. In a study conducted by Serra et al. [69], hydroalcoholic extracts of *P. venusta* stems presented high potential for the inhibition of ACE. According to the colorimetric method proposed, there is good correlation between the traditional methodology and the one proposed by this study for the prospection of ACE inhibitors. The *P. venusta* extracts tested presented 31.6% and 33.1% inhibition when evaluated by the traditional and proposed methodology, respectively. These results are very promising since Jabeen and Aslam [70] found 88.93 ± 3.78% ACE inhibitory activity by captopril at 2 μM concentration. It is noteworthy that captopril (D-3-mercapto-2-methyl-propionyl-L-proline) is a potent, competitive inhibitor of ACE, which is widely used orally as an antihypertensive agent [71].

REFERENCES

[1] Lohmann, L. G., Ulloa, C. U. (2006). *Bignoniaceae*. In: *iPlants prototype checklist*. http://www.iplants.org. Accessed 26 November 2018.

[2] APG (Angiosperm Phylogenetic Group) II. (2003). An update of the Angiosperm Phylogeny Group classification for the orders and families of flowering plants. *Bot. J. Linn. Soc.*, 141, 399-436.

[3] APG (Angiosperm Phylogenetic Group) III. (2009). An update of the Angiosperm Phylogeny Group classification for the orders and families of flowering plants. *Bot. J. Linn. Soc.*, 161, 105-121.

[4] Gentry, A. H. (1980). *Bignoniaceae*: part I (Crescentia and Tourrettieae). In: *Flora Neotropica Monograph*, vol 25(I). NYBG Press, New York.

[5] Lohmann, L. G. (2004). *Bignoniaceae*. In: Smith, N., Mori, S., Henderson, A., Stevenson, D., Held, S. (eds). *Flowering Plants of the Neotropics*. Princeton University Press, NYBG, New York.

[6] Fischer, E., Theisen, I., Lohmann, L. G. (2004). *Bignoniaceae*. In: Kubitzki, K., Kadereit, J. W. (eds). *The Families and Genera of Vascular Plants*. VII. Flowering plants. Dicotyledons. Lamiales (except Acanthaceae including Avicenniaceae). Springer, Berlin, pp 9-38.

[7] Olmstead, R. G., Zjhra, M. L., Lohmann, L. G., Grose, S. O., Eckert, A. J. (2009). A molecular phylogeny and classification of *Bignoniaceae*. *Amer. J. Bot.*, 96, 1731-1743.

[8] Grose, S. O., Olmstead, R. G. (2007). Evolution of a charismatic neotropical clade: molecular phylogeny of *Tabebuia* s.l., Crescentieae, and allied genera (*Bignoniaceae*). *Syst. Bot.*, 32, 650-659.

[9] Pace, M. R., Lohmann, L. G., Olmstead, R. G., Angyalossy, V. (2015). Wood anatomy of major *Bignoniaceae* clades. *Plant Syst. Evol.*, 301, 967-995.

[10] Pool, A. A. (2008). Review of the genus *Pyrostegia* (*Bignoniaceae*). *Ann. Mo. Bot. Gard.*, 95(3), 495-510.

[11] Milani, M., Heintze, W., Schafer, G., de Souza, P. V. D. (2015). Influência do número de folíolos e nós no enraizamento de estacas semilenhosas de cipó-de-são-joão. [Influence of leaf number and nodes on the rooting of semiwoody cuttings of flame vine]. *Ornam. Hortic.*, 21(3), 339-344.

[12] Gilman, E. F. (1999). *Pyrostegia venusta*. Fact Sheet FPS-496. University of Florida. Cooperative Extension Service. Institute of Food and Agricultural Sciences.

[13] Nisha, P. V., Shruti, N., Swamy, K. S., Kumari, M., Vedamurthy, A. B., Krishna, V., Hoskeri, J. H. (2012). Anthelmintic activity of *Pyrostegia venusta* using *Pheretima posthuma*. *Int. J. Pharm. Sci. Drug Res.*, 4(3), 205-208.

[14] Menninger, E. A. (1970). *Flowering Vines of the World*. Hearthside Press Inc., New York.

[15] Sampaio, E. S., Almeida, A. A. (1994). Influência da temperatura e luminosidade na floração de *Pyrostegia venusta* (*Bignoniaceae*) na região urbana de Curitiba, Paraná. [Floral morphology and

reproductive biology of *Pyrostegia venusta* (*Bignoniaceae*) in the urban region of Curitiba, Paraná]. *Acta Biol. Par.*, 23, 79-88.

[16] Singh, S., Rana, A., Chauhan, S. V. S. (2009). Impact of environmental changes on the reproductive biology in *Pyrostegia venusta* Presl. *J. Environ. Biol.*, 30(2), 271-273.

[17] Ferreira, D. T., Alvares, P. S. M., Houghton, P. J., Braz-Filho, R. (2000). Constituintes químicos das raízes de *Pyrostegia venusta* e considerações sobre a sua importância medicinal. [Chemical constituents from roots of *Pyrostegia venusta* and considerations about its medicinal importance]. *Quím. Nova*, 23(1), 42-46.

[18] Rodrigues, V. E. G., Carvalho, D. A. (2001). Levantamento etnobotânico de plantas medicinais no domínio do cerrado na região do Alto Rio Grande – Minas Gerais. [Ethnobotanical survey of medicinal plants in the region of the Alto Rio Grande – Minas Gerais]. *Ciênc. Agrotec.*, 25(1), 102-123.

[19] Lorenzi, H. (2008). *Plantas Daninhas do Brasil*. [*Weeds of Brazil*]. 4th ed. Plantarum, Nova Odessa.

[20] Scalon, S. P. Q., Vieira, M. C., Lima, A. A., Souza, C. M., Mussury, R. M. (2008). Tratamentos pré-germinativos e temperaturas de incubação na germinação de cipó-de-São-João (*Pyrostegia venusta* (Ker Gawl.) Miers) – *Bignoniaceae*. [Pregerminative treatments and incubation temperatures on the germination of “cipó-de-São-João” (*Pyrostegia venusta* (Ker Gawl.) Miers) - *Bignoniaceae*]. *Rev. Bras. Pl. Med.*, 10(4), 37-42.

[21] Sens, S. L. (2002). *Alternativas para a auto-sustentabilidade dos Xokleng da terra indígena Ibirama, Florianópolis*. [*Alternatives for the self-sustainability of the Xokleng of the Ibirama indigenous land, Florianópolis*]. Masters Dissertation, Universidade Federal de Santa Catarina.

[22] Harvey, A. L. (2008). Natural products in drug discovery. *Drug Discovery Today*, 13(19/20), 894-901.

[23] Dubey, R., Misra, K. J. (1976). Chemical components of *Pyrostegia venusta* flowers. *J. Indian Chem. Soc.*, 53, 378.

[24] Harbone, J. B. (1967). Comparative biochemistry of the flavonoids. VI. Flavonoid patterns in the *Bignoniaceae* and Gesneriaceae. *Phytochemistry*, 6, 1646-1651.

[25] Krishna, V., Sharma, S., Pareek, R. B., Singh, P. (2002). Terpenoid constituents from some indigenous plants. *J. Indian Chem. Soc.*, 79(6), 550-552.

[26] Gusman, A. B., Gottsberger, G. (1996). Differences in floral morphology, floral nectar constituents, carotenoids, and flavonoids in petals of orange and yellow *Pyrostegia venusta* (*Bignoniaceae*) flowers. *Phyton*, 36(2), 161-171.

[27] Tinoi, J., Rakariyatham, N., Deming, R. L. (2006). Determination of major carotenoid constituents in petal extracts of eight selected flowering plants in the north of Thailand. *Chiang Mai J. Sci.*, 33(2), 327-334.

[28] Bouzada, M. L. M., Fabri, L. R., Nogueira, M., Konno, T. U. P., Duarte, G. G., Scio, E. (2009). Antibacterial, cytotoxic and phytochemical screening of some traditional medicinal plants in Brazil. *Pharm. Biol.*, 47(1), 44-52.

[29] Roy, P., Amdekar, S., Kumar, A., Singh, V. (2011). Preliminary study of the antioxidant properties of flowers and roots of *Pyrostegia venusta* (Ker Gawl) Miers. *BMC Complement. Altern. Med.*, 11, 69.

[30] Veloso, C. C., Cabral, L. D. M., Bitencourt, A. D., Franqui, L. S., Santa-Cecília, F. V., Dias, D. F., Soncini, R., Vilela, F. C., Giusti-Paiva, A. (2012). Anti-inflammatory and antinociceptive effects of the hydroethanolic extract of the flowers of *Pyrostegia venusta* in mice. *Braz. J. Pharmacog.*, 22, 162-168.

[31] Moreira, C. G., Horinouchi, C. D. S., Souza-Filho, C. S., Campos, F. R., Barison, A., Cabrini, D. A., Otuki, M. F. (2012). Hyperpigmentant activity of leaves and flowers extracts of *Pyrostegia venusta* on murine B16F10 melanoma. *J. Ethnopharmacol.*, 141, 1005-1011.

[32] Pereira, A. M. S., Hernandes, C., Pereira, S. I. V., Bertoni, B. W., França, S. C., Pereira, P. S., Taleb-Contini, S. H. (2014). Evaluation of anticandidal and antioxidant activities of phenolic compounds

from *Pyrostegia venusta* (Ker Gawl.) Miers. *Chem.-Biol. Interact.*, 224, 136-141.

[33] Daneluzzi, G. S., Dos Santos, V. H. M., Silva, L. P., Da Silva, R. M. G. (2014). Avaliação dos potenciais fitotóxico e citotóxico de *Pyrostegia venusta* (Ker Gawl.) Miers (*Bignoniaceae*). [Evaluation of phytotoxic and cytotoxic potential of *Pyrostegia venusta* (Ker Gawl.) Miers (*Bignoniaceae*)]. *Biosci. J.*, 30(4), 1231-1240.

[34] Santos, M. D., Blatt, C. T. T. (1998). Teor de flavonóides e fenóis totais em folhas de *Pyrostegia venusta* Miers. de mata e de cerrado. [A quantitative analysis of foliar flavonoids and total phenolics of *Pyrostegia venusta* Miers. from both forest and cerrado]. *Braz. J. Bot.*, 21(2), 135-140.

[35] Blatt, C. T. T., Santos, M. D., Salatino, A. (1998). Flavonoids of *Bignoniaceae* from the "cerrado" and their possible taxonomic significance. *Pl. Syst. Evol.*, 210, 289-292.

[36] Altoé, T. D. P., Amorim, G. M., Gomes, J. V. D., Borges, A. S., Valadão, I. C., Silva, I. V., Rangel, L. B. A., Vieira, P. C., Jamal, C. M., Kitagawa, R. R., Borges, W. S. (2014). *In vitro* antioxidant and cell viability of *Pyrostegia venusta* (Ker Gawl.) Miers). *Orbital: Electron. J. Chem.*, 6(4), 215-222.

[37] Coimbra, M. C., Chagas, R. C. R., Duarte-Almeida, J. M., Castro, A. H. F. (2017). Influence of plant growth regulators and light on callus induction and bioactive phenolic compounds production in *Pyrostegia venusta* (*Bignoniaceae*). *Indian J. Exp. Biol.*, 55, 584-590.

[38] Dinda, B., De, U. C., Bhattacharya, A., Arima, S., Sato, N., Harigaya, Y. (2002). Chemical constituents of *Argyreia argetea*, *Millingtonia hortensis* and *Pyrostegia venusta*. *J. Indian Chem. Soc.*, 79(3), 291-293.

[39] Pinho, R. S., Oliveira, A. F. M., Silva, S. I. (2009). Potential oilseed crops from the semiarid region of northeastern Brazil. *Bioresour. Technol.*, 100, 6114-6117.

[40] Loredo-Carrillo, S. E., Santos-Díaz, M. L., Leyva, E., Santos-Díaz, M. S. (2013). Establishment of callus from *Pyrostegia venusta* (Ker

Gawl.) Miers and effect of abiotic stress on flavonoids and sterols accumulation. *J. Plant Biochem. Biotechnol.*, 22(3), 312-318.

[41] Braga, K. Q., Coimbra, M. C.; Castro, A. H. F. (2015). *In vitro* germination, callus induction and phenolic compounds contents from *Pyrostegia venusta* (Ker Gawl.) Miers. *Acta Sci., Biol. Sci.*, 37(2), 151-158.

[42] von Poser, G. L., Schripsema, J., Henriques, A. T., Jensen, S. R. (2000). The distribution of iridoids in *Bignoniaceae*. *Biochem. Syst. Ecol.*, 28, 351-366.

[43] Fischer, R., Vasilev, N., Twyamn, R. M., Schillberg, S. (2015). High-value products from plants: the challenges of process optimization. *Curr. Opin. Biotechnol.*, 32, 156-162.

[44] Muthy, H. N., Lee, E. J. Paek, K. Y. (2014). Production of secondary metabolites from cell and organ cultures: strategies and approaches for biomass improvement and metabolite accumulation. *Plant Cell Tiss. Organ Cult.*, 118, 1-16.

[45] Hussain, M. S., Fareed, S., Ansari, S., Rahman, M. A., Ahmad, I. Z., Saeed, M. (2012). Current approaches toward production of secondary plant metabolites. *J. Pharm. Bioallied. Sci.*, 4(1), 10-20.

[46] Oyebanji, O. B., Nweke, O., Odebunmi, O., Galadima, N. B., Idris, M. S., Nnodi, U. N., Afolabi, A. S., Ogbadu, G. H. (2009). Simple, effective and economical explant surface sterilization protocol for cowpea, rice and sorghum seeds. *Afr. J. Biotechnol.*, 8, 5395-5399.

[47] Coimbra, M. C., Castro, A. H. F. (2016). Different methods for surface sterilization of *Pyrostegia venusta* (Ker Gawl.) Miers (*Bignoniaceae*) leaf explants. *Plant Cell Cult. Micropropag.*, 12(2), 34-38.

[48] Rossatto, D. R., Kolb, R. M. (2010). Germinação de *Pyrostegia venusta* (*Bignoniaceae*), viabilidade de sementes e desenvolvimento pós-seminal. [Germination of *Pyrostegia venusta* (*Bignoniaceae*) seed viability and post-seminal development]. *Rev. Bras. Bot.*, 33(1), 51-60.

[49] Crispim, J. G., Rêgo, M. M., Rêgo, E. R., Pessoa A. M. S., Barrosa, P. A. (2012). Utilização de diferentes substratos na propagação de

Pyrostegia venusta através de estacas. [Utilization of different substrates in the propagation of *Pyrostegia venusta* by cuttings]. *Rev. Verde*, 10(4), 38-41.

[50] Engelmann, F. (2011). Use of biotechnologies for the conservation of plant biodiversity. *In Vitro Cell. Dev. Plant*, 47, 5-16.

[51] Smith, R. H. (2012). *Plant tissue culture: techniques and experiments*. Academic Press, San Diego.

[52] Torres, C. A., Zampini, I. C., Nuñes, M. B., Isla, M. I., Castro, M. P., Gonzalez, A. M. (2013). *In vitro* antimicrobial activity of 20 selected climber species from the *Bignoniaceae* family. *Nat. Prod. Res.*, 27(22), 2144-2148.

[53] Fernandes, A. P., Ribeiro, G. E., Rufino, L. R. A., Da Silva, L. M., Boriollo, M. F. G., Oliveira, N. M. S., Fiorini, J. E. (2011). Efeito do extrato hidroalcoólico de *Pyrostegia venusta* na mutagênese "*in vivo*," e avaliação antimicrobiana, e interferência no crescimento e diferenciação celular "*in vitro*". [Effect of the hydroalcoholic extract of *Pyrostegia venusta* on *in vitro* mutagenesis, antimicrobial assessment, *in vitro* growth and cell differentiation]. *Rev. Med. Minas Gerais*, 21(3), 272-279.

[54] Silva, P. B., Medeiros, A. C. M., Duarte, M. C. T., Ruiz, A. L. T. G., Kolb, R. M., Frei, F., Santos, C. (2011). Avaliação do potencial alelopático, atividade antimicrobiana e antioxidante dos extratos orgânicos das folhas de *Pyrostegia venusta* (Ker Gawl.) Miers (*Bignoniaceae*). [Evaluation of allelopathic potential, antimicrobial and antioxidant activities of *Pyrostegia venusta* (Ker Gawl.) Miers (*Bignoniaceae*) leaf organic extracts]. *Rev. Bras. Pl. Med.*, 13(4), 447-455.

[55] De Souza, M. B., Silva Júnior, J. O. C., Barbosa, W. L. R., Valério, E. S., Lima, A. M., Araújo, M. H., Muzitano, M. F., Nakamura, C. V., De Mello, J. C. P., Teixeira, F. M. (2016). *Pyrostegia venusta* (Ker Gawl.) Miers crude extract and fractions: prevention of dental biofilm formation and immunomodulatory capacity. *Pharmacogn. Mag.*, 12(2), S218-S222.

[56] Roy, P., Amdekar, S., Kumar, A., Singh, R., Sharma, P., Singh, V. (2012). *In vivo* antioxidative property, antimicrobial and wound healing activity of flower extracts of *Pyrostegia venusta* (Ker Gawl) Miers. *J. Ethnopharmacol.*, 140, 186-192.

[57] Brandão, G. C., Kroon, E. G., Dos Santos, J. R., Stehmann, J. R., Lombardi, J. A., De Oliveira, A. B. (2010). Antiviral activity of *Bignoniaceae* species occurring in the State of Minas Gerais (Brazil): part 1. *Lett. Appl. Microbiol.*, 51, 469-476.

[58] Veloso, C. C., Oliveira, M. C., Oliveira, C. C., Rodrigues, V. G., Giusti-Paiva, A., Teixeira M. M., Duarte, I. D., Ferreira, A. V. M., Perez, A. C. (2014). Hydroethanolic extract of *Pyrostegia venusta* (Ker Gawl.) Miers flowers improves inflammatory and metabolic dysfunction induced by high-refined carbohydrate diet. *J. Ethnopharmacol.*, 151, 722-728.

[59] Moreira, C. G., Carrenho, L. Z. B., Pawloski, P. L., Soley, B. S., Cabrini, D. A., Otuki, M. F. (2015). Pre-clinical evidences of *Pyrostegia venusta* in the treatment of vitiligo. *J. Ethnopharmacol.*, 168, 315-325.

[60] Veloso, C. C., Bitencourt, A. D., Cabral, L. D. M., Franqui, L. S., Dias, D. F., Dos Santos, M. H., Soncini, R., Giusti-Paiva, A. (2010). *Pyrostegia venusta* attenuate the sickness behavior induced by lipopolysaccharide in mice. *J. Ethnopharmacol.*, 132, 355-358.

[61] Da Silva, R. M. G., Dos Santos, V. H. M., Daneluzzi, G. S., Silva, L. P. (2012). Potencial alelopático de *Pyrostegia venusta* (*Bignoniaceae*) sobre a germinação e crescimento radicular de couve-da-malásia. [Allelopathic potential of *Pyrostegia venusta* (*Bignoniaceae*) on the germination and root growth of *Brassica chinensis* L. var. *parachinensis*]. *Rev. Agricult.*, 87(3), 139-150.

[62] Gianfaldoni, S., Wollina, U., Tirant, M., Tchernev, G., Lotti, J., Satolli, F., Rovesti, M., França, K. Lotti, T. (2018). Herbal compounds for the treatment of vitiligo: a review. *Open Access Maced. J. Med. Sci.*, 6(1), 203-207.

[63] Silva, R. M. G., Rodrigues, D. T. M., Augustos, F. S., Valadares, F., Neto, P. O., Santos, L., Silva, L. P. (2012). Antitumor and cytotoxic

activity of *Kielmeyera coriacea* Mart. Zucc. and *Pyrostegia venusta* (Ker-Gawl.) Miers extracts. *J. Med. Plants Res.*, 6(24), 4142-4148.

[64] Figueiredo, C. R., Matsuo, A. L., Pereira, F. V., Rabaça, A. N., Farias, C. F., Girola, N., Massaoka, M. H., Azevedo, R. A., Scutti, J. A. B., Arruda, D. C., Silva, L. P., Rodrigues, E. G., Lago, J. H. G., Travassos, L. R., Da Silva, R. M. G. (2014). *Pyrostegia venusta* heptane extract containing saturated aliphatic hydrocarbons induces apoptosis on B16F10-Nex2 melanoma cells and displays antitumor activity *in vivo*. *Pharmacogn. Mag.*, 10(38), S363-S376.

[65] Magalhães, E. A., Silva Júnior, G. J., Campos, T. A., Silva, L. P., Da Silva, R. M. G. (2010). Avaliação do potencial genotóxico do extrato bruto de *Pyrostegia venusta* (Ker Gawl.) Miers, Bignoneaceae, em medula óssea de camundongos. [The Evaluation of the genotoxic potency of the *Pyrostegia venusta* (Ker Gawl.) Miers, Bignoneaceae, crude extract on bone marrow of mice]. *Braz. J. Pharmacog.*, 20(1), 65-69.

[66] Gomes, M. D., Teixeira, C. A., Barbosa, G. M., Froes, M. T., Silva, M. B. Da Veiga, V. F., Soares, R. M., Dos Santos, A. L., Holandino, C. (2012). Effects of direct electric current on *Herpetomonas samuelpessoai*: an ultrastructural study. *Bioelectromagnetics*, 33, 334-345.

[67] Mori, Y., Nakazawa, M., Tomimatsu, H., Momma, K. (2000). Long-term effect of angiotensin-converting enzyme inhibitor in volume overloaded heart during growth: a controlled pilot study. *J. Am. Coll. Cardiol.*, 36(1), 270-275.

[68] Yimgang, D. P., Sorkin, J. D., Evans C. F., Abrahan, D. S., Rosenthal, G. L. (2017). Angiotensin converting enzyme inhibitors and interstage failure in infants with hypoplastic left heart syndrome. *Congenit. Heart Dis.*, 13, 533-540.

[69] Serra, C. P., Côrtes, S. F., Lombardi, J. A., Braga de Oliveira, A., Braga, F. C. (2005). Validation of a colorimetric assay for the *in vitro* screening of inhibitors of angiotensin-converting enzyme (ACE) from plant extracts. *Phytomedicine*, 12, 424-432.

[70] Jabeen, Q., Aslam, N. (2013). Hypotensive, angiotensin converting enzyme (ACE) inhibitory and diuretic activities of the aqueous-methanol extract of *Ipomoea reniformis*. *Iran J. Pharm. Res.*, 12(4), 769-776.

[71] Odaka, C., Mizuochi, T. (2000). Angiotensin-converting enzyme inhibitor captopril prevents activation-induced apoptosis by interfering with T cell activation signals. *Clin. Exp. Immunol.*, 121(3), 515-522.

In: Phytochemicals
Editor: Iman Ryan
ISBN: 978-1-53615-478-8

Chapter 11

LIMONIASTRUM FEEI FROM THE ALGERIAN SAHARA: ETHNOPHARMACOLOGY, PHYTOCHEMISTRY AND THE PHARMACOLOGICAL IMPORTANCE

Abdelkrim Cheriti*
Phytochemistry and Organic Synthesis Laboratory
Faculty of Medicine, UTMB, Algeria

Chapter dedicated to the memory of my father Cheriti Bouhafs who died on 25 November 2017

ABSTRACT

Traditional remedies have been employed for the treatment and management of various ailments since the beginning of human civilization. The plant family *Plumbaginaceae* comprises a large number

*Corresponding Author's E-mail: karimcheriti@yahoo.com.

of species (836 species grouped in 27 genera) that have been and are still used as medicinal plants, particularly in folk medicine.

In the Algerian Sahara, the *Plumbaginaceae* family is represented only by two genera *Limonium* and *Limoniastrum.*

The medicinal specie *Limoniastrum feei* (Girard) Batt., is endemic to Algeria and Morocco growing in arid conditions has been used ethnobotanically for treatment of a variety of disorders, bronchitis, diarrhea, gastrointestinal tracts, as anti-inflammatory, for skin diseases, treatment of infected wounds, hepatic pains, children fever and in addition the plant provides good grazing for camel.

The aim of this chapter is to present as much information as was established from the available scientific literature. The review comprises the ethnopharmacological, biological activities and phytochemical of the *Limoniastrum feei* specie growing in the Algerian Sahara and used as medicinal plants.

Keywords: *Plumbaginaceae*, *Limoniastrum feei*, ethnopharmacological, biological activities, phytochemical, Sahara

INTRODUCTION

Medicinal plants have been successfully applied in clinics and as an important source of pharmacological substances around the world. One of the most efficient ways of finding new bioactive compounds is collecting data on the use of medicinal plants in traditional pharmacopeia. Historically, archeological studies have indicated that medicinal plants were regularly employed by people in prehistoric times for at least 60,000 years. In several ancient cultures, botanical products were ingested for biomedically curative and psychotherapeutic purposes. Plants have formed the basis of Traditional Medicine (TM) systems that have been in existence for thousands of years and continue to provide mankind with new remedies, such as, the oldest known medicinal systems of the world: Ayurveda, Arabian medicine, Chinese and Kempo medicine [1-6]. Natural products have been of great importance in disease treatment since ancient timesand have made enormous contributions to human health through compounds such as quinine, morphine, aspirin, digitoxin and many others, so natural products are so important to undertake research are that they can be a source of new compounds. Thus, the World Health Organization (WHO) has recognized the potential utility of traditional remedies and strives to preserve primary health care involving medicinal plants. Nearly 50,000 species of higher plants have been used for medicinal purposes. Natural products and their derivatives represent more than 50% of all the drugs in clinical use in the world and in which higher plants contribute to no less than 25% [7]. Molecular biology is also being used in this process and the pharmacological profiles of these compounds are screened using new research equipment and new technology. In several part of the world, only a small part of the 400.000 vegetable species known were investigated from phytochemical and pharmacological aspects, and that each species can contain up to several thousands of different components [8,9]. It is widely known that the presence of Chemodiversity of bioactive compounds is responsible for the pharmacological effects of medicinal plants.

The plant family *Plumbaginaceae* comprises a large number of species that have been and are still used as medicinal plants, particularly in folk medicine (mostly *Limoniastrum*, *Limonium* and *Plumbago* species) as stomachic, for treating diarrhea, gastrointestinal tracts, as anti-inflammatory, for skin diseases and used as a food and as garden ornaments [10]. Additionally, crude extracts of some species from *Plumbaginaceae* family have been reported to exhibit antibacterial, antiparasitic, antiviral, antioxidant, anti-inflammatory, immunomodulatory, cytotoxic, antitumor, anti-amyloidogenic and molluscicidal activities. From a chemical point of view, different secondary metabolites including terpenoids, triterpenoid saponin, sterol, chalcone, tannin, coumarins, flavonoids, flavones glycosides and phenolic compounds have been reported [11, 12].

Algeria with its large area and diversified climate has a varied flora about 4450 taxa of which 3950 indigenous with 6.5% endemic [13]. This richness flora is a source of rich and abundant medical matter and, in particular, Sahara part constitutes an important reservoir of many plants which have not been investigated until today. The *Plumbaginaceae* family is represented in the flora of Algeria by 33 species, among this flora, the endemic specie *Limoniastrum feei,* widely used in the Algerian Sahara ethnopharmacopea as a medicinal plant for the treatment of various diseases [14]. We attempt to present a review on the study of ethnopharmacological, biological activities, and phytochemical analysis of plants from the genus *Limoniastrum*, especially the endemic *Limoniastrum feei* growing in the Algerian Sahara and used as medicinal plants.

BOTANICAL TAXONOMY OF THE GENUS *LIMONIASTRUM*

Plumbaginaceae (Limoniaceae, Juss. 1789, Dicots.) is a nearly cosmopolitan family that consists of 27 genera and about 836 species, whereas a later study on Caryophyllales identified 29 genera for the family [15]. The total number of species ascribed to these genera ranges from one (e.g., *Bamiania* Lincz., *Bukiniczia* Lincz., *Ghaznianthus* Lincz., *Saharanthus* M.B. Crespo & M.D. Lledó) to 596 (*Limonium)* [16]; making *Plumbaginaceae* one of the top 20% of angiosperm families in terms of species richness [17]. The center of diversity is in the Mediterranean area and central and western parts of Asia [18, 19]. Common name of the *Plumbaginaceae* is leadworts, from the Latin words plumbum ("lead") and agere ("to resemble"). Species of *Plumbaginaceae* are mostly perennial; they are in a form of shrubs, subshrubs, lianas or herbs. The leaves are alternately arranged, in basal rosettes, or in fascicular rosettes on annual shoots below scapes. Foliage is simple with entire margin, rarely pinnatifid. Usually leaves are with short or without petiole. The flowers are regular, hypogynous and pentamerous. Each flower consists of a synsepalous calyx, a corolla that is usually sympetalous, five antepetalous stamens and a superior pentacarpellary unilocular gynoecium with one basal ovule. Based on morphological and chemical features *Plumbaginaceae* family was divided in two marked subfamilies Plumbagoideae (Plumbaginoideae) and Armerioideae (Limonioideae or Staticoideae [18-20]. APG II and APG IV (Angiosperm Phylogeny Group II and IV system) of plant classification that was published in 2003 and 2016 by the Angiosperm Phylogeny Group, placed *Plumbaginaceae* in the order Caryophyllales Perleb and the family is sister to Polygonaceae Juss. Hence, the most recent systematic classification of the *Plumbaginaceae* family is given below [21]:

Kingdom:	Plantae
Subkingdom:	Tracheobionta
Division:	Magnoliophyta Cronquist, Takht. & Zimmerm. ex Reveal
Class:	Magnoliopsida Brongn.
Subclass:	Caryophyllidae Takht.
Order:	Caryophyllales Perleb
Family:	*Plumbaginaceae* Juss.

Almost all over the 27 genera of *Plumbaginaceae* growing in arid and maritime habitats are as follows: *Acantholimon, Aegialitis, Afrolimon, Armeria, Bakerolimon, Bamiania, Bukiniczia, Cephalorrhizum, Ceratostigma, Chaetolimon, Dictyolimon, Dyerophytum, Eremolimon, Ghaznianthus, Gladiolimon, Goniolimon, Ikonnikovia, Limoniastrum, Limoniopsis, Limonium, Muellerolimon, Neogontscharovia, Plumbagella, Plumbago, Popoviolimon, Psylliostachys, Vasssilczenkoa* [18].

In the flora of Algeria, *Plumbaginaceae* family is represented by 4 genus containing 33 species divided on: *Arméria Willd* (6 species), *Plumbago Tourn* (1 specie), *Limoniastrum Moench* (3 species) and *Limonium Tourn* (23 species) [22].

The genus *Limoniastrum Moench* (Family: *Plumbaginaceae* Juss, subfamily Staticoideae Kostel) is shrubs, found mostly in coastal and saline dry areas of the Mediterranean and sub-Saharan Africa [23]. As described by Kubitzki [18], Limoniastrum genera: Dwarf shrubs. Leaves narrow. Inner bract enveloping the flowers. Calyx tubular, at margin broadened. Corolla tube about as long as the lobes. Filaments basally sometimes bearded. Styles connate up to the midle. Stigmas filiform-cylindric. This genera comprise seven species including: *L. weygandiorum, L. rechingeri, L. migiurtinum, L.monopetalum, L. ifniens, L. Guyonianum* and *Limoniastrum feei*. Most of these species are medium-sized shrubs, commonly living in coastal salt marshes or saline and dry regions. Morphologically they are distinct from other genera of *Plumbaginaceae* in having petals that form a long tube, convolute inner bracts and styles that are connate for half their length [18]. Limoniastrum genera has been divided into two sections or subgenera, Limoniastrum (Eulimoniastrum) for tall shrubs with alternate leaves and 3-bracteate spikelets with a hornless inner bract and sect. Bubania Batt. for dwarf shrubs with rosulate leaves and 2-bracteatespikelets with a horned inner bract [24].

The species of Limoniastrum genus have an important role in ecosystems and play a very effective role in the stabilization of coastal dunes. Moreover, the formation of galls in the shoots of these species constitutes an ecological habitat of a large number of insects such as Lepidoptera and Hymenoptera (*Oecococecis guyonella, Acalyptris limoniastri*) [25].

Figure 1. The endemic *Limoniastrum feei* - Algeria Sahara.

Limoniastrum feei [syn. *Ceratolimon feei, Bubania feei*], vernacular name: *Melfet khadem* (POSL Herbarium CA99/14), which is endemic to south west Algeria and south east morocco, and used in the Algerian Sahara ethnopharmacopea as medicinal plants [14] (Figure 1).

Limoniastrum feei is perennial shrub with densely ramified stems at the base, rather small, not exceeding 40 cm. The leaves form a basal rosette; they are lanceolated, approximately 5 cm long, slightly fleshy, and covered with small calcareous concretions. Solitary flowers develop at the tips of the long thin stems. They have a thick reddish calyx and 5 pink petals. The fruits are small capsules. This Saharan specie thrives particulary on the stony grounds of the djebels. It frequently occurs on gravelly sand wadi beds and can survive in severe drought when the rainfall is less than 50 mm per year [22, 23].

ETHNOPHARMACOLOGY AND BIOACTIVITY OF THE GENUS *LIMONIASTRUM*

The plant family *Plumbaginaceae* comprises a large number of species that have been and are still used as medicinal plants, particularly in folk medicine (mostly *Limoniastrum*, *Limonium* and *Plumbago* species) as stomachic, for treating diarrhea, gastrointestinal tracts, as anti-inflammatory, for skin diseases and used as a food and as garden ornaments [10]. Additionally, crude extracts of some species from *Plumbaginaceae* family have been reported in the literature justifying some of the medicinal uses of the species. These pharmacological activities include antibacterial, antiparasitic, antiviral, antioxidant, anti-inflammatory, immunomodulatory, cytotoxic, antitumor, anti-amyloidogenic, and molluscicidal activities. Among three species of the *Limoniastrum* genus (*L. feei, L. guyonianum* and *L. monopetalum*) growing in Algeria (North Africa), two species have been evaluated for their pharmacological activities, following their uses in traditional ethnopharmacopeae.

Limoniastrum feei (Vernacular name: *Melefet khadem*) commonly used in popular medicine especially in Sahara as an antidiarrhoic, antispasmodic, to cure sore throats, respiratory ailments cough, children fever, hepatitis and in the treatment of gastric disorders, skin diseases and infected wounds. Preparations are usually made by using the aerial parts of the plant boiled in hot water. The plant is appreciated by livestock, mainly by camel [4, 14, 26]. The second specie *Limoniastrum guyonianum* (Vernacular name: *Zita*) has been used in traditional medicines to treat dysentery and gastric infections especially against infectious diseases or parasites responsible for painful and bloody diarrhea. The galls of this species are used for tanning leather and dyeing hair [10, 27-29].

In the last years, the search for new antibiotics has accelerated, due to increasing frequency of multidrug-resistant bacteria. Many plants have been used for their antimicrobial properties due to their bioactive constituents, including phenolic compounds, terpenoids, saponins, alkaloids and essential oils [30]. Data on pharmacological evaluation of *Limoniastrum* species are scarce. Preliminary observations were published in 2005 by team of Phytochemistry and Organic Synthesis Laboratory, thus, in our first study of *Limoniastrum feei*, we have studied the antimicrobial evaluation by disc diffusion method of several crude extracts of leaves, stem, and twig parts of this specie against fungi (*Candida albicans, Saccharomyces cerevisiae*) and bacteria (*Escherichia coli, Entercocus fecalis, Staphylococcus aureus, Pseudomonas aeruginosa* and *Klebsiella pneumonia*). We have found that methanol extract from leaves of this endemic specie contained potential anti-fungal constituents that could be employed against *Candida albicans* and anti-bacterial constituents useful against *E. coli* [31]. Further, it was reported that the essential oil extracted from *L. guyonianum* leaves, flowers, seeds and roots exhibited antimicrobial activity [32].

As a part of our work on medicinal plants of the Algerian Sahara, we have reported that the Hydroalcoholic extract from aerial parts of *Limoniastrum feei* was evaluated for acute and subacute toxicity in *Swiss* mice after ingestions of the extract. The LD50 of the extract is higher than 1.53 g/kg and the subacute treatment did not shows any change in corporal weight and haematological parameters, suggest that the plant seems to be destitute of toxic effects in mice [33]. Aerial part and roots of

Limoniastrum feei were used to evaluate their extracts for antifungal activity against *Fusarium oxysporum* f. sp. albedinis Foa. The antifungal test was conducted using disc diffusion technique and relative virulence (RV) test (on potato tuber tissue). For both tests, four extract quantities were used (200, 400, 800 and 1,600µg). The relative virulence was presented as necrotic tissue weight (mg) of potato tuber tissue. Among all solvents, methanol had the best extraction yield (mean: 6.35%, minimum: 2.27%, maximum: 9.80%). The highest frequency of antifungal effect on Foa was presented by ethyl acetate extracts with a best values of zones of inhibition (µ: 19 mm) of *Limoniastrum feei* (aerial part, 1600µg) [34].

Recently, in our research on the molluscicidal plant extracts, we have found that aqueous extract of *Limoniastrum feei* exhibited a highest molluscicidal activity against *Lymnaea stagnalis. Thus,* the lethal concentrations for aqueous extracts of *Limoniastrum feei* (Leaves stem and twig) that killed 50% (LC50) of adult *Lymnaea stagnalis* were 74.10, 52.78 and 72.94 µg/ml, respectively, while the respective LC90 values were 91.81, 66.48 and 90.06 µg/ml [35]. On another side, in Algeria, the Saharan termite, *Anacanthotermes ochraceus* is among the most devastating termite pests, it cause great damage to house hold materials and agriculture in oasis. In our research on bioinsecticides from Saharan plants, among plants tested as termiticidal and eco-friendly termite control agents, aqueous extracts from aerial part of *Limoniastrum feei* showed moderate anti-termitic activity against termite workers in a different doses (lethal effect appears after 3h with 50% mortality for concentrations of 3%, 4% and 5%,), the observed activity is less than found in tested toxic saharan plants (*Calotropis procera, Hyoscyamus muticus, Pergularia tomentosa* and *Datura stramonium*) [36, 37].

It is well known that antioxidants entered in public's nutritional vocabulary in the 1990s, at a time when researchers were slowly discovering how oxygen-triggered free radical reactions in the body play a key role in aging-associated chronic diseases [38]. Currently, an antioxidant is defined as any substance able to eliminate ROS and derivatives (RNS, or reactive sulfur species, RSS), directly or indirectly, acting as an antioxidant defense regulator, or reactive species production inhibitor [39]. By using two methods (chemical and electrochemical), recently in our laboratory, Keffous et al., [40] were investigated the total flavonoids (TF), total phenolics, and antioxidant activity of *Limoniastrum feei* aqueous extract. Thus, the results show that *Limoniastrum feei* contains 200.28 μg of total phenolic in 1 mg of dry extract, expressed as gallic acid equivalents. The TF represent 54.77 μg/mg expressed as quercetin equivalents. The antioxidant activity of extracts has been evaluated by chemical and electrochemical methods. In the reducing power and total antioxidant capacity tests, the antioxidant activity of extracts was expressed as Ascorbic acid equivalents where the aqueous extract has an equivalent capacity of 233.39 and 112.4 μg for 1 mg, respectively. In $DPPH^{\bullet}$ radical trapping test, the IC 50 was equal to 0.58 mg/ml. The cyclic voltammetry of the extract indicates one oxidation irreversible peak at approximately 300–320 mV/(Ag/AgCl). The superoxide scavenging assay of *Limoniastrum feei* aqueous extract showed an average activity of order 61.46% at 0.5 mg/ml doses.

In their study to determine the in vitro antioxidant activities of *Limoniastrum guyonianum* roots and their cytoprotective action on H_2O_2-challenged rat small intestine epithelial cells (IEC-6 cells), Bettaib et al., [41] demonstrated that *L. guyonianum* at a low dose may have an important role in redox homeostasis control within H_2O_2-challenged IEC-6 cells without being toxic to the control ones. The activity at low dose occurs by enhancing CAT activity, increasing GSH levels, and decreasing JNK phosphorylation to efficiently prevent cell death.

In comparative study on two *Limoniastrum* species, Debouba et al., [42] indicat that *L. guyonianum* and *L. monopetalum* have similar free radical DPPH (2,2-diphenyl-1-picrylhydrazyl) scavenging activity, *L. guyonianum* showed more than two-fold higher ferrous ions chelating activity relative to *L. monopetalum*. However, highest reducing power activity was observed in *L. monopetalum*. Thiobarbituric acid-reactive substances (TBARS) determination indicated that *L. monopetalum* behave better lipid membrane integrity relative to *L. guyonianum*. Therefore, the authors suggested that the lesser stressful state of *L. monopetalum* was related to higher metabolites accumulation and reducing capacity compared to *L. guyonianum*. An important antioxidant and mutagenic effect of *Limoniastrum guyonianum* gall extract were observed by testing the aqueous extract on *Salmonella typhimurium*, the result showed no mutagenicity when tested with *S. typhimurium* strain TA104 either with or without exogenous metabolic activation mixture (S9). The study revealed that the aqueous gall extract induced a mutagenic effect in *S. typhimurium* TA1538 when tested in the presence, as well as in the absence, of S9 activation mixture at the concentration of 500 mg/mL [43]. The authors obtained a significant inhibition percentage (76.39%) toward 3.25 mg/plate of methylmethanesulfonate. Antimutagenicity against aflatoxin B1, 4-nitro-o-phenylene-diamine and 2-aminoanthrace`ne was significant, with an inhibition percentage of, respectively, 70.63, 99.3 and 63.37% in the presence of, respectively, *S. typhimurium* TA100, *S. typhimurium* TA1538 and *S. typhimurium* TA104 strains at a concentration of 250 mg/plate after metabolic activation (S9). In continuation of their works on *Limoniastrum guyonianum* aqueous gall extract, Krifa et al., [44] demonstrated that aqueous gall extract may act as chemotherapeutic agents, inducing inhibition of the growth and apoptosis in BE colorectal cancer cells, and proposed an apoptotic mechanism in which calpain inhibition may participate in the apoptosis induced by aqueous gall extract in BE cells, So the extract from *Limoniastrum guyonianum* gall can be an interesting way to prevent and/or to treat colorectal cancer by inhibiting calapin activity. Previous investigations revealed that *Limoniastrum guyonianum* aqueous gall extract was able to induce splenocyte proliferation and to stimulate macrophage activation [45]. The ethyl acetate extract of this specie contained epigallocatechingallate which exhibited antitumor effects that were discovered from various cancer cell lines, animal models and clinical studies [28, 46]. In the same way, a study was done to evaluate the anti-cancer effect of *L. guyonianum* aqueous gall extract and luteolin in the human cervical cancer HeLa cell line, the results show that gall extract induces apoptosis related to a reduced DNA methylation likely by acting on the epigenetic integrator UHRF1 and inhibits cell proliferation and arrests cell cycle progression at the G2/M phase which could be through re-expression the tumor suppressor gene p16INK4A [27]. A study was conducted to investigate possible antiproliferative activity of *Limoniastrum feei*. Thus, *n*-BuOH fraction from the ethanolic-aqueous extract (80%) of *L. feei* whole plant was evaluated against the cell line: J774.A1, murine monocyte/macrophage, WEHI-164, murine fibrosarcoma, and HEK-293, human epithelial kidney cells. The extract is active at IC50 = 27 µg/mL [47].

Inflammation is the immune system's response to harmful stimuli, such as pathogens, damaged cells, toxic compounds, or irradiation [48], and acts by removing injurious stimuli and initiating the healing process [49]. Various pathogenic factors, such as infection, tissue injury, or cardiac infarction, can induce inflammation by causing tissue damage. Inflammation is frequently a key element in the pathological progression of organ disease. It is estimated that some 15% of human cancers are associated with chronic infection and inflammation [50, 51]. Recently we have demonstrated the potential analgesic properties of crude aqueous extract of leaves from Limoniastrum feei. Thus, we have evaluated for analgesic activity the extract by using acetic acid induced writhing in mice. The extract at the tested doses 25 and 30mg/kg produced a percentage inhibition of the acetic acid induced abdominal constriction of 33.39 and 35.38% respectively. The crude extract produced a statistically significant analgesic activity comparable to the effect of standard drug (50mg/kg Diclofenac) [52]. Also in another study, we noted that the dose of 25mg/kg of the aqueous leaves extract of *Limoniastrum feei* significantly reduced the paw edema of mouse induced by formalin in a reduction of inflammation relative to diclofenac, of 85% and 95% [53].

PHYTOCHEMISTRY OF THE GENUS *LIMONIASTRUM*

Medicinal and aromatic plants have demonstrated significant importance in primary health care and are therefore highly significant in the process of drug development and discovery. Plants secondary metabolites, commonly referred to as phytochemical or biologically active compounds that exhibit a holistic healing approach.

Table 1. Natural compounds isolated from *Limonium* species

Class of compounds	Isolated natural compounds (N°)	Source (Specie) [Ref.]
Flavanone	Naringenin **1**, Eriodictyol **2**, Homoeriodictyol **3**	*Limonium aureum* [60], *Limonium bonduelli* [61], *Limonium sinens* [62]
	Homoeriodictyol-*7-O-β*-D-glucopyranoside **4**	*Limonium aureum* [61]
Flavonol	Myricetin **5**	*Limonium aureum* [61], *Limonium axillare* [64], *Limoniumdensiflorum* [63],*Limonium gmelinii* [64],*Limonium myrianthum [65], Limonium popovii* [64], *Limonium sinense* [62]
	Quercetin **6**	*Limonium* cv. ('Gold Coast') [66], *Limonium aureum* [67], *Limonium axillare* [64],*Limonium gmelinii* [64], *Limonium popovii* [64], *Limonium sinense* [62]
	Kaempferol **7**	*Limonium aureum* [67], *Limonium axillare* [64, 68], *Limonium gmelinii* [64], *Limonium popovii* [64]
	Myricetin-*3-O-β*-glucopyranoside **8**	*Limonium aureum* [67], *Limonium gmelinii* [64], *Limonium popovii* [64], *Limonium wrightii* [69]
	Myricetin-*3-O-β*-galactopyranoside **9**	*Limonium aureum* [67], *Limonium gmelinii* [64], *Limonium popovii* [64], *Limonium sinense* [62], *Limonium tetragonum* [70],
	Myricetin-*3-O-α*-L-rhamnopyranoside **10**	*Limonium* cv. ('Gold Coast') [66], *Limonium aureum* [67], *Limonium brasilience* [71], *Limonium gmelinii* [64], *Limonium popovii* [64], *Limonium sinense* [[62], *Limonium tetragonum* [70], *Limonium wrightii* [69]
	Myricetin-*7-O-α*-L-rhamnopyranoside **11**	*Limonium aureum* [67], *Limonium gmelinii* [64], *Limonium popovii* [64],

Table 1. (Continued)

Class of compounds	Isolated natural compounds (N°)	Source (Specie) [Ref.]
Flavonol	Myricetin-*3*-*O*-*α*-L-arabinopyranoside **12**	*Limonium aureum* [67], *Limonium gmelinii* [64],*Limonium popovii* [64], *Limonium sinense* [62], *Limonium wrightii* [69]
	Myricetin-*3*-*O*-*β*-L-xylopyranoside **13**	*Limonium gmelinii* [64], *Limonium popovii* [64]
	Myricetin-*3*-*O*-*β*-D- sorboside **14**	*Limonium axillare* [71]
	Myricetin-*3*-*O*-*α*-L-rhamnopyranosyl-(1-6)-*β*-D-glucopyranoside **15**	*Limonium gmelinii*[64], *Limonium myrianthum [65], Limonium popovii*[64]
	Myricetin-*3*-*O*-(6″-galloyl-*β*-glucopyranoside) **16**	*Limonium aureum* [67], *Limonium gmelinii* [64], *Limonium popovii* [64]
	Myricetin-*3*-*O*-(6″-galloyl-*β*-galactopyranoside) **17**	*Limonium aureum* [67], *Limonium gmelinii* [64], *Limonium popovii* [64], *Limonium sinense* [62]
	Myricetin-*3*-*O*-*β*-D-(2″-galloylarabinopyranoside) **18**	*Limonium gmelinii* [64], *Limonium popovii* [64],
	Myricetin 3-*O*-(2″-galloylrhamnopyranoside) **19**	*Limonium sinense* [62]
	Myricetin 3-*O*-(3″-galloylrhamnopyranoside) **20**	*Limonium sinense* [62]
	Myricetin 3-*O*-(4″-galloylrhamnopyranoside) **21**	*Limonium sinense* [62]
	Myricetin 3-*O*-(2″-*p*- hydroxybenzoylrhamnoside) **22**	*Limonium sinense* [62]
	Myricetin-3'-methylether-*7*-*O*-*β*-D-glucopyranoside **23**	*Limonium Sinuatum* [72]
	Quercetin-*3*- *O*-*β*-D-glucopyranoside **24**	*Limonium wrightii* [69], *Limonium tetragonum* [70],
	Quercetin-*3*-*O*-*β*-D-galactopyranoside **25**	*Limonium aureum* [67], *Limonium tetragonum* [70], *Limonium wrightii* [69]

Class of compounds	Isolated natural compounds (N°)	Source (Specie) [Ref.]
Flavonol	Quercetin -3-*O*-rhamnopyranoside **26**	*Limonium* cv. ('Gold Coast') [66], *Limonium aureum* [67], *Limonium sinense* [62]
	Quercetin 3-*O*-arabinopyranoside **27**	*Limonium wrightii* [69]
	Rutin (quercetin-3-O-rutinoside) **28**	*Limonium gmelinii* [64], *Limonium popovii* [64]
	Kaempferol 3-*O*-glucopyranoside **29**	*Limonium wrightii* [69]
	Isorhamnetin-7-*O*-*β*-D-glucopyranoside **30**	*Limonium Sinuatum* [72]
	Gmelinoside I **31**	*Limonium gmelinii* [64,65]
Flavone	Luteolin **32**	*Limonium* cv. ('American Beauty. 'Blue Bonnet', 'Iceberg', 'Midnight Blue', 'Twilight Lavender') [73], *Limonium axillare* [74], *Limonium bonduelli* [61], *Limonium sinense* [62]
	Luteoline-7-*O*-glucopyranoside **33**	*Limonium axillare* [74]
	Apigenin **34**	*Limonium* cv. ('Gold Coast') [66, 73], *Limonium axillare* [74], *Limonium bonduelli* [61]
	Apiine **35**	*Limonium axillare* [74]
C-Glycosyl flavone	Isoorientin **36**	*Limonium* cv. ('American Beauty. 'Blue Bonnet', 'Iceberg', 'Midnight Blue', 'Twilight Lavender', 'Gold Coast') [66, 73]
Catechin-gallate	(-)-Epigallocatechine **37**, (+)-Gallocatechine **38,** (-)-Epigallocatechine-3-*O*-gallate **39,** (−)-Epigallocatechin 3-(3-methoxygallate **40** (−)-Epigallocatechin 3-(3,5-dimethoxygallate) **41**	*Limonium brasilience* [71], *Limonium gmelinii* [75], *Limonium popovii* [64], *Limonium sinense* [62]

Table 1. (Continued)

Class of compounds	Isolated natural compounds (N°)	Source (Specie) [Ref.]
Anthocyanin	Delphinidin 3-*O*-glucopyranoside **42**	*Limonium* cv. ('American Beauty', 'Blue Bonnet', 'Twilight Lavender', 'Iceberg') [73], *Limonium brasilience* [71]
	Delphinidine-3-rhamnopyranoside **43**, Delphinidine-3,5-di-glucopyranoside **44**	*Limonium brasilience* [71]
Aurone	4,6,4′-Trihydroxyaurone **45**, Aureusidin **46**, Aureusidin 4-*O*-glucopyranoside **47**, Aureusidin 6-*O*-glucopyranoside **48**	*Limonium* cv.('Gold Coast') [66]
Tannin	**Samarangenin B 49**	*Limonium sinense* [68]
	Geraniin **50**	*Limoniumaxillare* [76]
Chalcone	3,4,2′,4′,6′-Pentahydroxychalcone **51**	*Limonium* cv. ('Gold Coast') [66]
Coumarin	Aesculétine **52**, Bergaptène **53**,Impératorine **54**, Isoimpératorine **55**, Ombelliferone **56**, Xanthotoxol **57**	*Limoniumaxillare* [76]
phénolic acids	Ellagic acid **58**, Ferulic acid **59**, Isoferulic acid **60,** Gallic acid **61**, **S**yringic acid **62**, Vanillic acid **63**	*Limoniumaxillare* [64,76], *Limonium bonduelli* [61], *Limonium brasilience* [71], *Limonium gmelinii* [64], *Limonium myrianthum [65], Limonium popovii* [64]
Sterol	Sitosterol **64**, Stigmasterol **65**, Sitosterol glucopyranoside **66**	*Limoniumaxillare* [71]
Essential oils	Main compounds: Hexacosane (10.7%),Palmitic acid (9.76%), Nonacosane (8.44%), (E,E)-Farnesyl acetate (7%), Vanilline (6.51%)	*Limoniumechioides* [77]

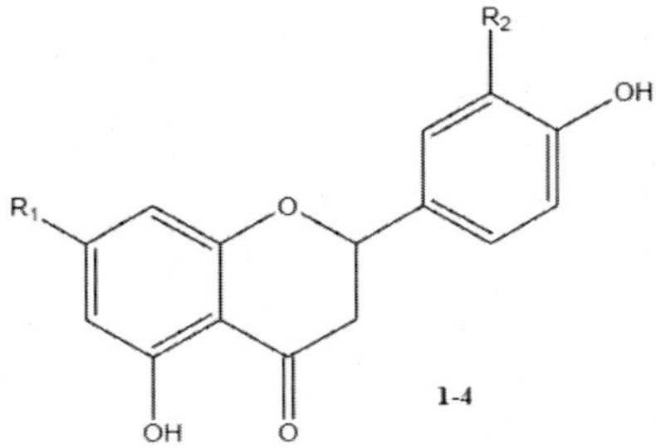

1-4

Comp. N°	R_1	R_2
1	OH	H
2	OH	OH
3	OH	OCH3
4	*O*-*β*-D-glu	OCH3

Figure 2. Chemical structures of Flavanone (**1-4**) from *Limonium* species.

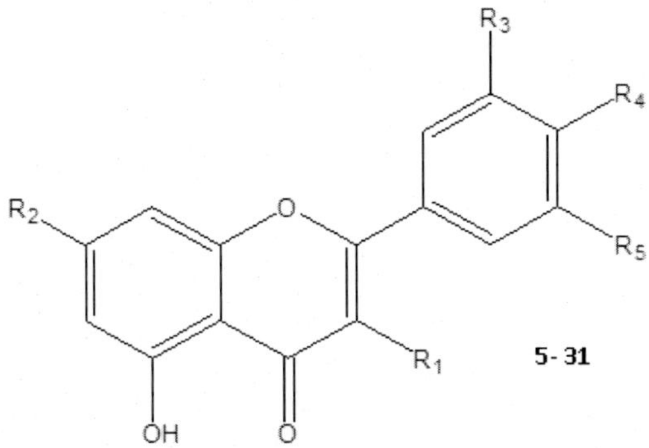

5- 31

Comp.N°	R_1	R_2	R_3	R_4	R_5
5	OH	OH	OH	OH	OH
6	OH	OH	H	OH	OH
7	OH	OH	H	OH	H
8	*O*-*β*-glu	OH	OH	OH	OH
9	*O*-*β*-gal	OH	OH	OH	OH
10	*O*-*α*-L-rha	OH	OH	OH	OH
11	OH	O-α-L-rha	OH	OH	OH
12	*O*-*α*-L-ara	OH	OH	OH	OH
13	*O*-*β*-L-xyl	OH	OH	OH	OH
14	*O*-*β*-D-sorboside	OH	OH	OH	OH

Figure 3. (Continued).

Comp.N°	R_1	R_2	R_3	R_4	R_5
15	*O*-*α*-L-rha-(1-6)-*β*-D-glu	OH	OH	OH	OH
16	*O*-(6''-galloyl)-*β*-glu	OH	OH	OH	OH
17	*O*-(6''-galloyl)-*β*-gal	OH	OH	OH	OH
18	*O*-*β*-D-(2''-galloyl)-ara	OH	OH	OH	OH
19	*O*-*β*-D-(2''-galloyl)-rha	OH	OH	OH	OH
20	*O*-*β*-D-(3''-galloyl)-rha	OH	OH	OH	OH
21	*O*-*β*-D-(4''-galloyl)-rha	OH	OH	OH	OH
22	*O*-(2″-O-p-hydroxybenzoyl)-*α*-rha	OH	H	OH	OH
23	OH	O-β-D-glu	H	OCH_3	OH
24	*O*-*β*-D-glu	OH	H	OH	OH
25	*O*-*β*-D-gal	OH	H	OH	OH
26	*O*-*α*-L-rha	OH	H	OH	OH
27	*O*-*α*-L-ara	OH	H	OH	OH
28	*O*-*α*-L-rut	OH	H	OH	OH
29	*O*-*β*-D-glu	OH	H	OH	H
30	OH	O-□-D-Glu	H	OCH_3	OH
31	3-O-α-D-gal	OH	OH	OH	OH

Figure 3. Chemical structures of Flavonol (**5-31**) from *Limonium* species.

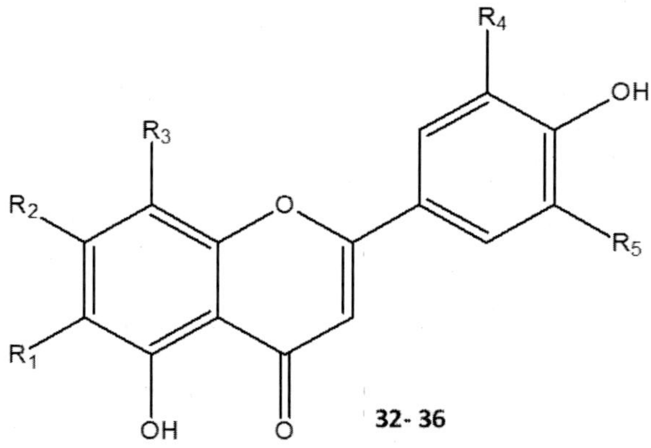

32- 36

Comp. N°	R_1	R_2	R_3	R_4	R_5
32	H	OH	H	OH	H
33	H	*O*- *β*-D-glu	H	OH	H
34	H	OH	H	H	H
35	H	O-[β-D-apiosyl-(1-2) -*β*-D-glu]	H	H	H
36	*O*- *β*-D-glu	OH	H	H	OH

Figure 4. Chemical structures of Flavone (**32-36**) from *Limonium* species.

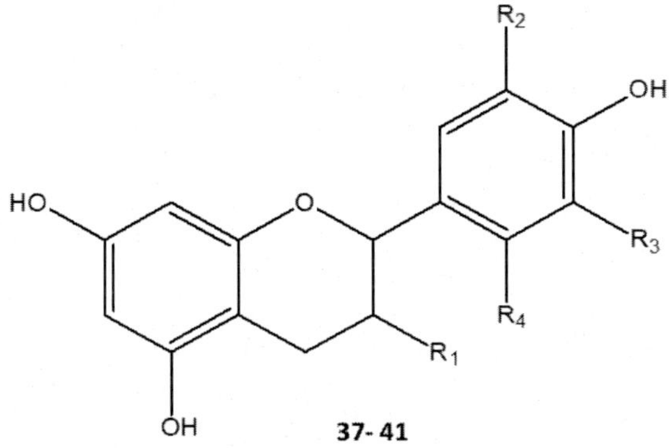

37- 41

Comp. N°	R_1	R_2	R_3	R_4
37	OH	H	OH	H
38	OH	OH	OH	H
39	O-gallate	OH	OH	H
40	O-(3-methoxy)gallate	H	OH	H
41	O-(3,5-dimethoxy)gallate	OH	OH	H

Figure 5. Chemical structures of Catechine (**37-41**) from *Limonium* species.

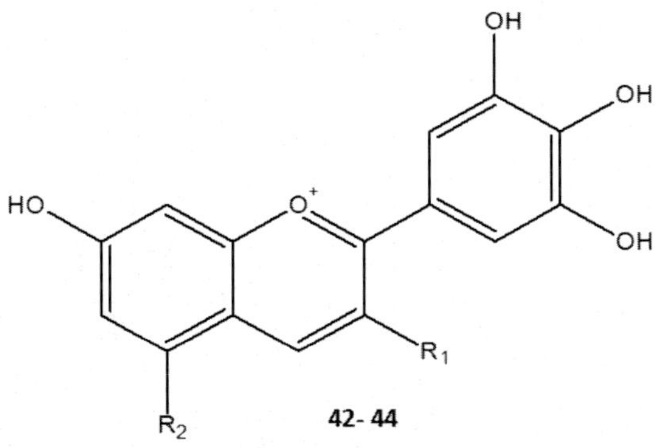

42- 44

Comp. N°	R_1	R_2
42	*O*-glu	OH
43	*O*-rha	OH
44	*O*-glu	*O*-glu

Figure 6. Chemical structures of Anthocyanine (**42-44**) from *Limonium* species.

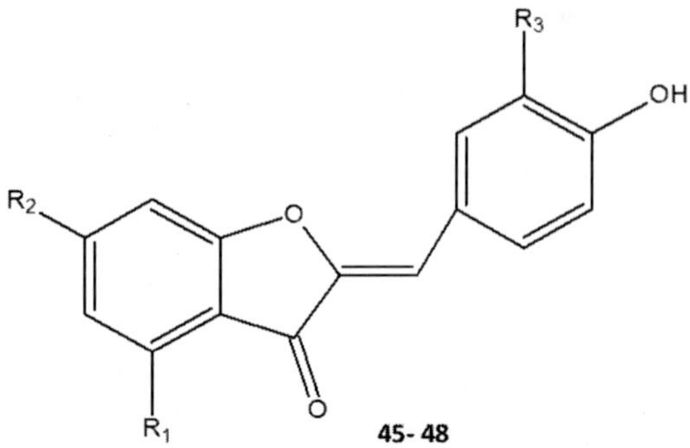

45- 48

Comp. N°	R_1	R_2	R_3
45	OH	OH	H
46	OH	OH	OH
47	*O*-glu	OH	OH
48	OH	*O*-glu	OH

Figure 7. Chemical structures of Aurone (**45-48**) from *Limonium* species.

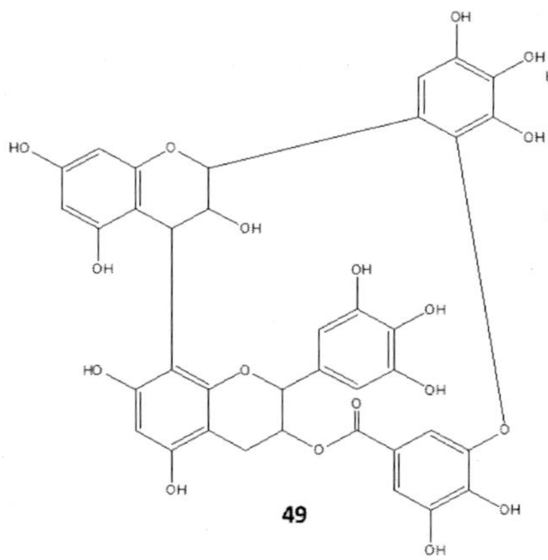

49

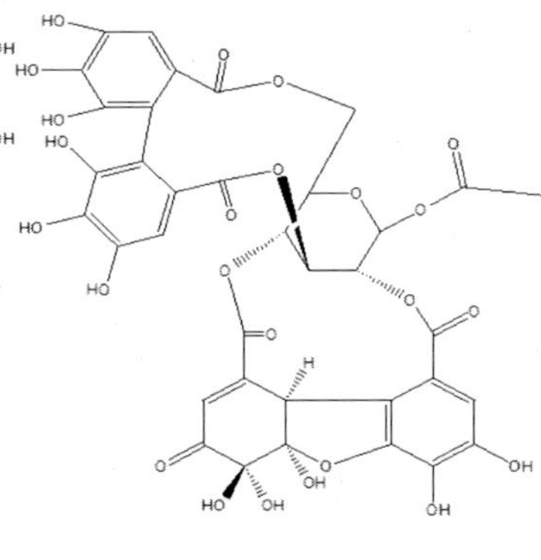

50

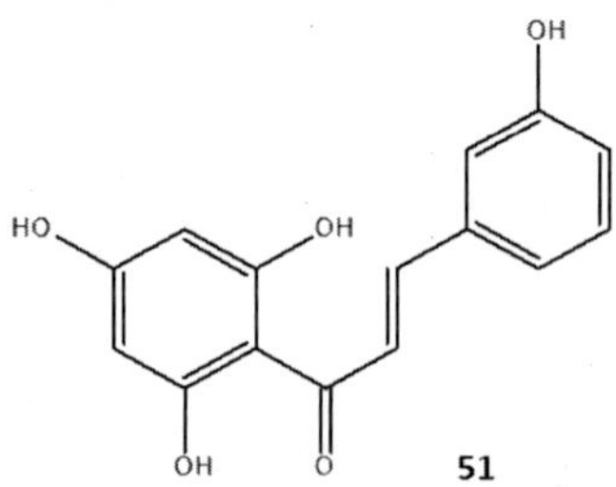

51

Figure 8. Chemical structures of Tannins **49, 50** and Chalcone **51** from *Limonium* species.

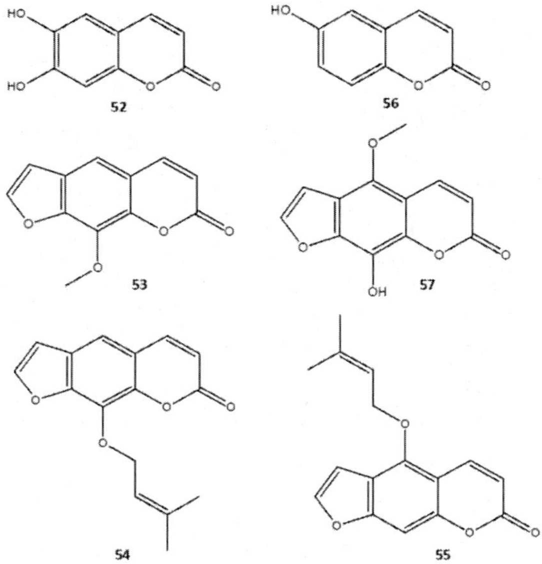

Figure 9. Chemical structures of Coumarines (**52-57**) from *Limonium* species.

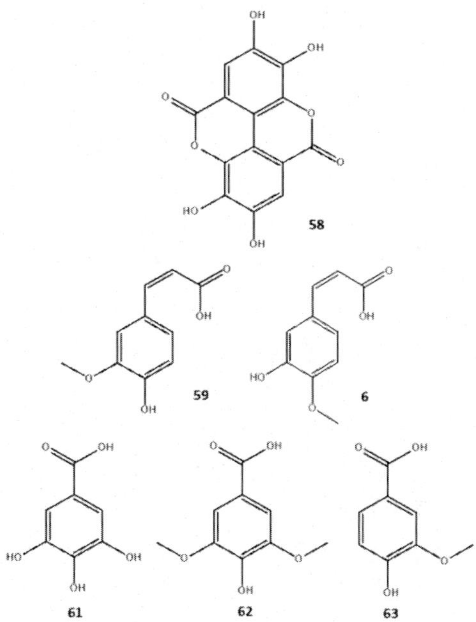

Figure 10. Chemical structures of Phenolic acids (**58-63**) from *Limonium* species.

64 R= H
66 R= β-D-Glu

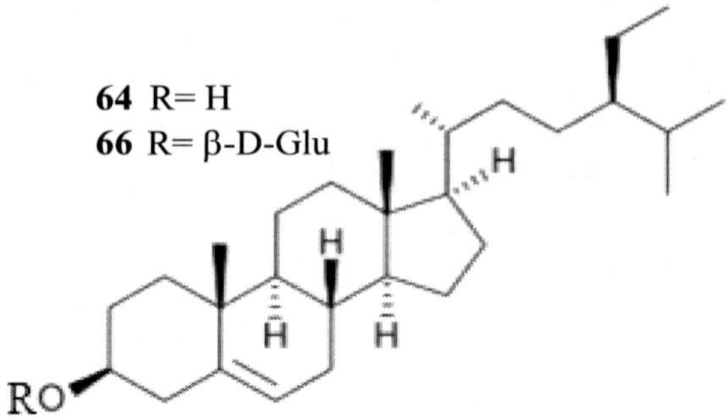

65

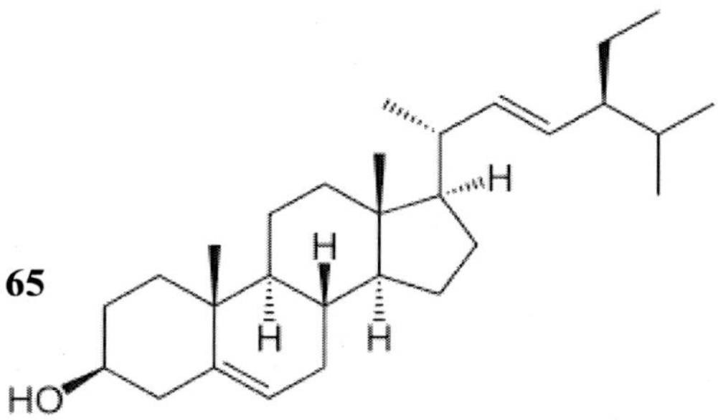

Figure 11. Chemical structures of Sitosterol (**64-66**) from *Limonium* species.

These include alkaloids, flavonoids, coumarins, quinones, saponins, terpenoids, tannins, glycosides, carbohydrates, proteins, phenolic compounds, essential oils, fixed oils and fats [1, 7, 8, 54]. Due to the high structural diversity and wide range of biological activities of their constituents [55], thus, a great number of medicinal species have been analyzed under the bioguided approach, and a compound or group of compounds have been isolated and tested as the biologically active agents that play the key role in the pharmacological effect [56, 57].

There is little phytochemistry works on the *Limoniastrum* genus, in a molecular study of *Plumbaginaceae* family, molecular results, together with morphological data, indicate that some studied *Limoniastrum* was sister to *Limonium* genus [58, 59]. Due to similarity of the *Limonium* genus with *Limoniastrum* and its importance as the most species rich genus of the *Plumbaginaceae* family with center of diversity is the Mediterranean region, where *ca.* 70% of all *Limonium* species are endemic[16]. It is interesting for us to give some information on the phytochemistry of this genus, which may have a similar chemotaxonomy as *Limoniastrum.* So far, many phytochemical studies of *Limonium* species have been conducted, and diversity natural compounds have been isolated and identified from species of this genus. Among them, flavonoides are the most abundant and major bioactive constituents in *Limonium*, and are considered as the characteristic constituents of this species. The identified compounds are listed in the following table and the corresponding structures are also comprehensively presented (Table 1, Figures 2-11).

Up to now, there are few reports on the phytochemical works on Limoniastrum genus, but we found some interesting works were various bioactive substances are reported especially from the two species: *Limoniastrum guyonianum* and *Limoniastrum feei*.

Chemical data on the endemic Saharan specie *Limoniastrum feei* are scarce in literature and few published papers describe phenolic components of this plant. In their bioguided fractionation of leaves extract studies in 2008, Chaabi et al. isolated common polyphenolic constituents such as myricetin **5**, quercetin **6**, myricetin-3-*O*-α-galactopyranoside **9**, myricetin 3-*O*-α-rhamnopyranoside **10,** epigallocatechin gallate **39**, gallic acid **61** and myrciaphenone A **67**. The authors demonstrate that the isolated gallic acid **61** was the most antioxidant compound in DPPH [(0.94 d 0.68) μg/mL] and FRAP [(0.83 d 0.15) μm $Fe^{2+/}$mL] tests, whereas myricetin **5** was the most active product in the superoxide nitroblue tetrazolium hypoxanthine /xanthine oxidase test [(1.86 d 0.12) μg/mL] [78].

In addition to the already isolated products, recently the same team were isolated from *n*-buthanol another compounds identified as salidroside **68** and a new natural compound with sulfate group characterized as 3,4-di hydroxyphenylthanol-8'-*O*-β-D-(6-sulpho)-glucopyranoside **69** [47].

In our first study in 2007 on the phytochemical investigation of this medicinal plant collected from the Sahara, we have described the isolation from the water-acetone extract of twig part of *Limoniastrum feei* four flavonoids. Two yellow amorphous powders identified as 6,3',4'-tri-methoxy 3,5,5'-trihydroxy flavonol **70**, 3-(6"-malonyl-2"-ramnosyl) - 6,3',4'-tri-methoxy 5,5'-dihydroxy flavonol **71** and a viscous gums identified as 7-dihydroxy-4'-Methoxy 8-O-β-glucopyranoside isoflavone **72** and 7,4'-diMethoxy 8-O-β-glucopyranoside isoflavone **73** [79].

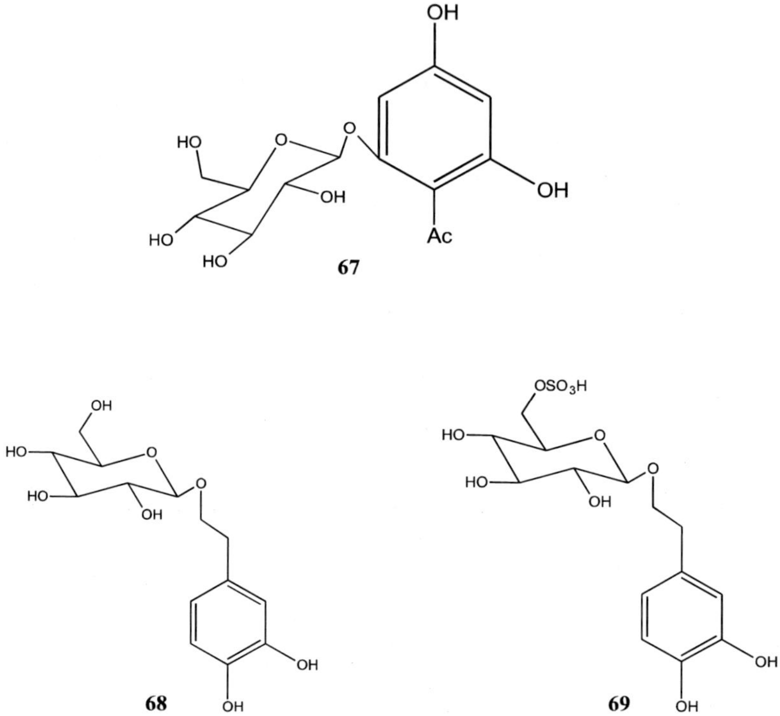

Figure 12. Chemical structures of phenolic components (**67-69**) from *L. feei.*

On the other hand, and in continuation of our investigation of this endemic medicinal plant, we have developed a liquid chromatography-UV method (LC-UV) for the analysis of the ethyl acetate fraction of the antibacterial methanolic extract of the aerial part of this species, we have characterized two major constituents identified as flavonoid glycosides - 6-dimethyl 2,5-hexyl-7-O-(1-6)diglucopyranoside-5,7,3,4-hydroxyl flavone **74** and 3-O-[6-(ester-3-Me diene 2,4 hexanoate)]-7-O-[diglucopyranosyle (1-6)(2-6)]-glucopyranoside-trihydroxyl 5,3,4-flavone **75** [80].

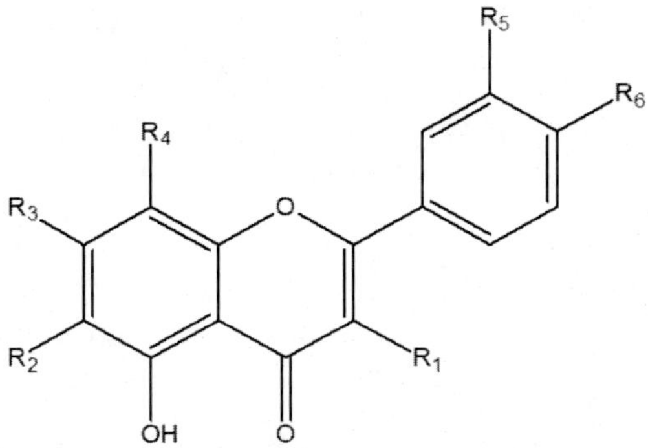

70, 71, 74, 75

	R_1	R_2	R_3	R_4	R_5	R_6
70	OH	OCH_3	OCH_3	OCH_3	OH	OH
71	*O*-(6”-malonyl-2”-rha)-β-glu	OCH_3	OCH_3	OCH_3	OH	OH
74	H	diMe-2,5-hexyl	OH	O-diglu-(1-6)	OH	OH
75	O-[6-(ester-3-Me diene2,4-hexanoate)]-glu	H	O-[diglu (1-6)(2-6)]-glu	H	OH	OH

Figure 13. Chemical structures of flavonol (**70**, **71**) and flavone (**74**,**75**) derivatives from *L. feei.*

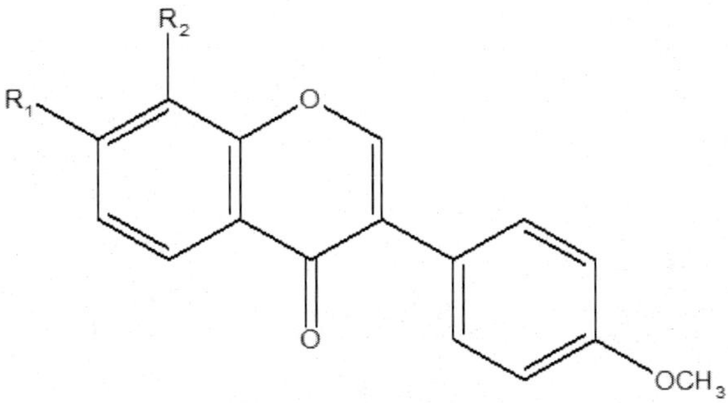

72, 73

	R_1	R_2
72	OH	*O*-glu
73	OCH_3	*O*-[α-rha-(1-6)]-β-glu

Figure 14. Chemical structures of isoflavone derivatives (**72**, **73**) from *L. feei.*

By application of the above technical LC-UV for the analysis and separation of saponin extracts from stems of *Limoniastrum feei.* Thus in first, the TLC analysis ($CHCl_3$:MeOH, 65:35) revealed four constituents in the butanol fraction (yield 0.60%, *Rf* 0.40, 0.48, 0.53, 0.58), six constituents in the ethyl acetate extract (yield 0.42%, *Rf* 0.48, 0.58, 0.63, 0.66, 0.78, 0.86), and five constituents in the methylene chloride extract (yield 0.31%, *Rf* 0.6, 0.66, 0.76, 0.80, 0.88). By using Si-gel phase liquid chromatography and gradient elution with $CHCl_3$: MeOH (65:35) as mobile phase, the LC-UV analysis for butanolic extract gave eight constituents, among them, two compounds were isolated and purified by flash chromatography (Si-gel column), eluting with H_2O/MeCN 75:35 and identified as 2α,3β,23-trihydroxy-3-O-acetylolean-12-ene **76**, and1-O-[α-L-rhamnopyranosyl-(1)-6-O-acetyl-β-D-galactopyranosyl]-1β,3β,22ξ-26-tetrahydroxyfurost-5(6)-ene-26-O-β-D-glucopyranoside **77** [81].

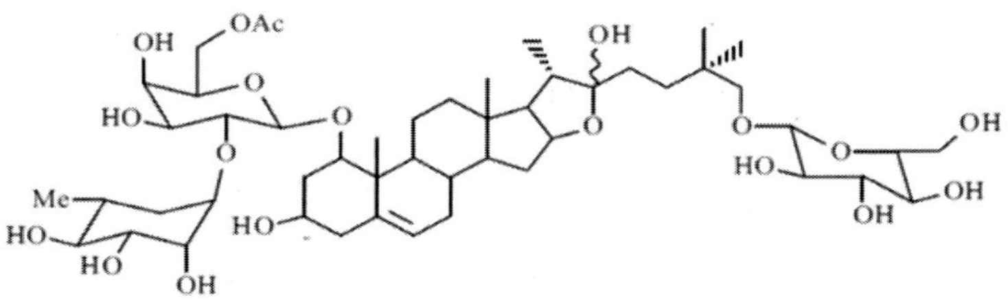

76

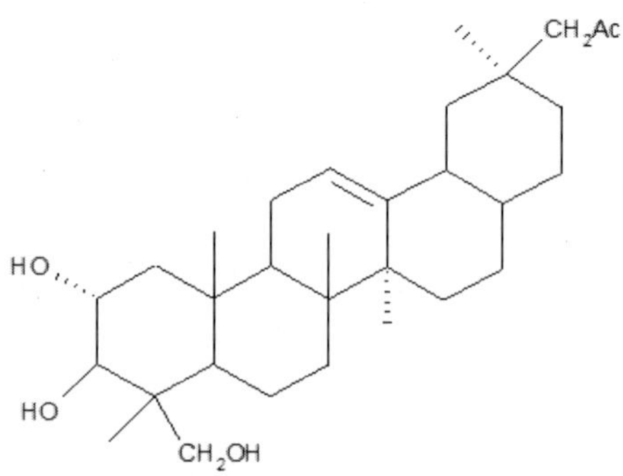

77

Figure 15. Chemical structures of triterpenoid derivatives (**76**, **77**) from *L. feei.*

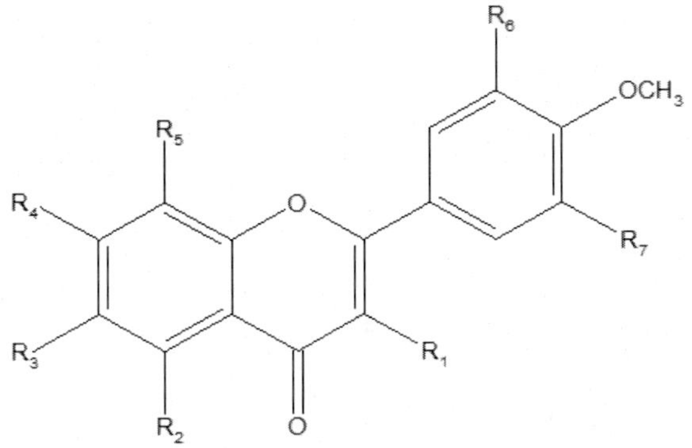

78-83

	R_1	R_2	R_3	R_4	R_5	R_6	R_7
78	H	OH	OCH_3	OCH_3	OCH_3	H	OCH_3
79	H	OCH_3	H	OH	H	H	OCH_3
80	H	OH	H	OH	OH	H	OCH_3
81	H	OH	OH	OH	H	OCH_3	OCH_3
82	*O*-(6”-malonyl-2”-rha)-β-glu	OH	H	OCH_3	H	OCH_3	OH
83	OH	OH	H	OCH_3	H	OCH_3	OH

Figure 16. Chemical structures of flavone derivatives (**78**, **83**) from twigs of *L. feei.*

In the same way, from the diethyl ether, ethyl acetate and *n*-butanol fractions of the acetone-water extract, we isolated by LC-UV eleven polyphenol compounds, distributed in the three organs as follows: six flavones derivatives (**78-83**) from twigs, four isoflavone derivatives (**72, 73, 84, 85**) from stems and one epigallocatechin gallate **39** from leaves [82].

The fractionation of acetone-water crude extract (diethyl ether, dichloromethane and ethyl acetate) of the aerial parts of *Limoniastrum feei* resulted in the isolation of seven compounds, which were characterized structurally as flavonoids, named: 5-hydroxy-3',4'-methoxyisoflavone **86**, 4'-methoxyisoflavone-7-O-β-glycopyranoside **87**, 3-hydroxy-5,6,7,4'-tetramethoxy flavone **88**, 5,4'-diméthoxy-3,6-dihydroxyflavones **89**, 7-O-(α-ramnopyranosyl-(1-6)-β-glycopyranosyl)-3-(3",4"-dimethyl-3"-

pentényl) 4', 5-dihydroxyflavones **90** and 6-C-β-(2"-O-β-glucopyranosyl-glucopyranosyl)-5,7,4'-trihydroxyflavones **91**[83].

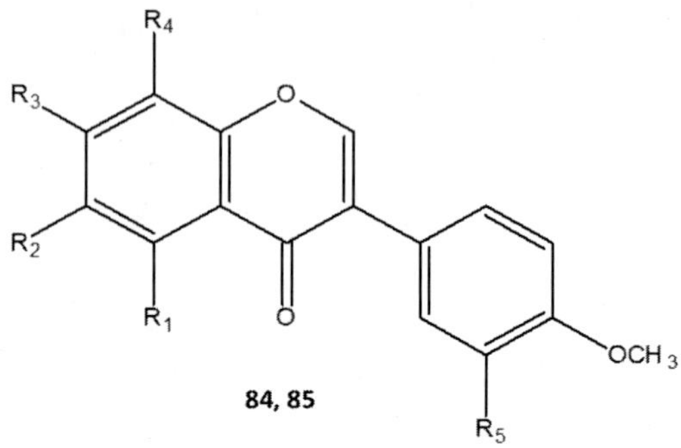

84, 85

	R_1	R_2	R_3	R_4	R_5
84	OH	OCH_3	OCH_3	OCH_3	OCH_3
85	OH	H	OH	H	OCH_3

Figure 17. Chemical structures of isoflavone derivatives (**84**, **85**) from stems of *L. feei.*

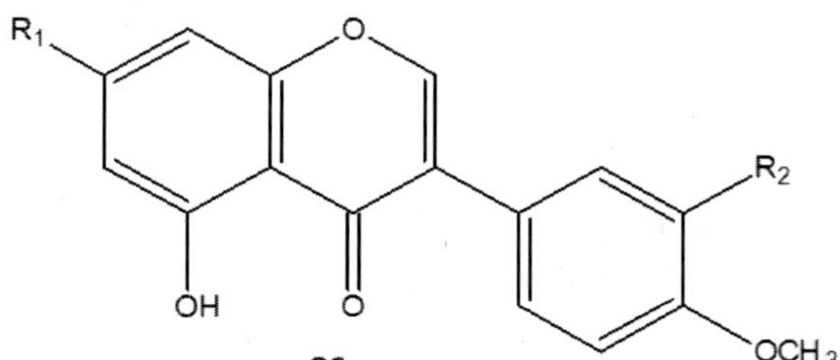

86, 87

	R_1	R_2
86	H	OCH_3
87	*O-β*-glu	H

Figure 18. Chemical structures of isoflavone derivatives (**86**, **87**) from the fractionation of acetone-water crude extract of the aerial parts of *L. feei.*

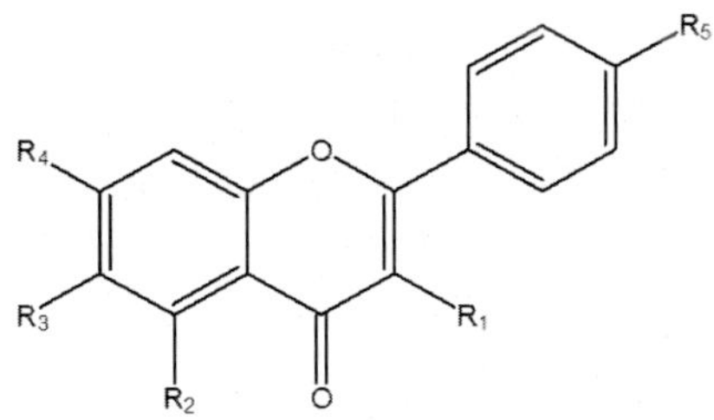

88-91

	R1	R2	R3	R4	R5
88	OH	OCH_3	OCH_3	OCH_3	OCH_3
89	OH	OCH_3	OH	H	OCH_3
90	3",4"diMe-3" pentenyl	OH	H	O-α-rha-(1-6)-β-glu	OH
91	H	OH	C-β- (2"-O-β – glu)-glu	OH	OH

Figure 19. Chemical structures of flavone derivatives (**88**, **91**) from the fractionation of acetone-water crude extract of the aerial parts of *L. feei.*

The hydrodistillation of the aeriel part of *Limoniastrum feei* gave brownish oil in a yield of 0.05% from dried material. By GC-MS analysis, twenty-seven compounds were identified, representing 95.8% of the total oil. The essential oil of *L. feei* was a mixture of different substances, including oxygenated monoterpenes (2%), aldehydes and Ketones (0.77%) and acids (0.47%). The aromatic ester as dominant group in the oil such as dioctyl phtalate (60.17%) and dibutyl phtalate (19.21%), we also found (R)-(-)-massoilactone in considerable amounts (4.55%) [84].

The medicinal specie from Tunisia, *Limoniastrum guyonianum* was studied for the first time for the essential oil obtained from different organs (leaves, flowers, seeds and roots). The chemical analysis by gas chromatography-flame ionization detector (GC-FID) and gas chromatography-mass spectroscopy (GC-MS), indicate the presence of different substances. Thus, in roots and leaves oils, monoterpenes were found in almost equal level (22.56 and 26.21%, respectively). (3Z) hex-3-enylmethanoate was detected as the chemotype in roots, while furfural (14.63%), methyl-2,4-dimethylbenzoate (14.70%) and 3-phenylprop-2 enylpentanoate (15.05%) were the most abundant constituents from seeds, leaves and flowers, respectively. The monoterpenol 2-(4-methylcyclohex-3-enyl) propanol was identified as common constituent of roots, leaves and flowers. The authors noted that the proportions of sesquiterpenes increased in the order stems (completely absent), roots (3%), flowers (7%) and leaves (10%), whereas the proportions of monoterpenes decreased from leaves to roots, flowers and stems (26, 23, 10 and 6%) [32].

In 2010, Trabelsi et al., by HPLC analysis of leaves extracts from *Limoniastrum monopetalum* using RP-HPLC coupled with UV-Vis multi-wave length detector identified seven phenolic acids: gallic **61** and vanillic **63** acids (as predominant phenolic compounds), ferulic **59**, syringic **62**, p-hydroxybenzoic, chlorogenic, and trans-cinnamic acids and four flavonoıds: quercetin **6**, apigenin **34**, amentoflavone **92** and flavone **93** [85].

The same team, in their investigation of the antioxidant capacities of the crude extract, aqueous and ethyl acetate partitions of *Limoniastrum guyonianum*, isolated three powerful compounds identified as epigallocatechin **37**, gallocatechin **38** and epigallocatechin-3-O-gallate **39**[86].

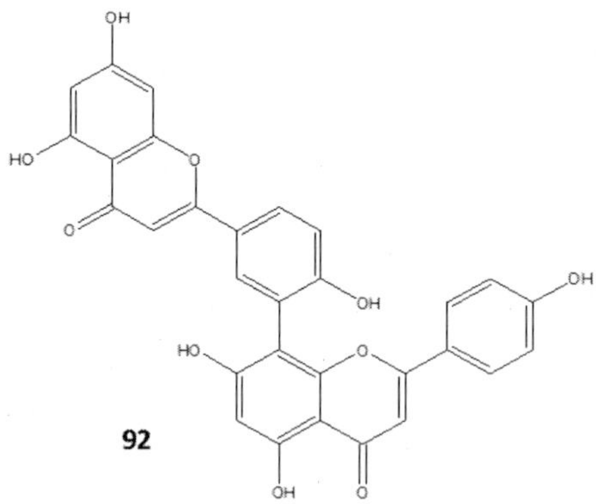

92

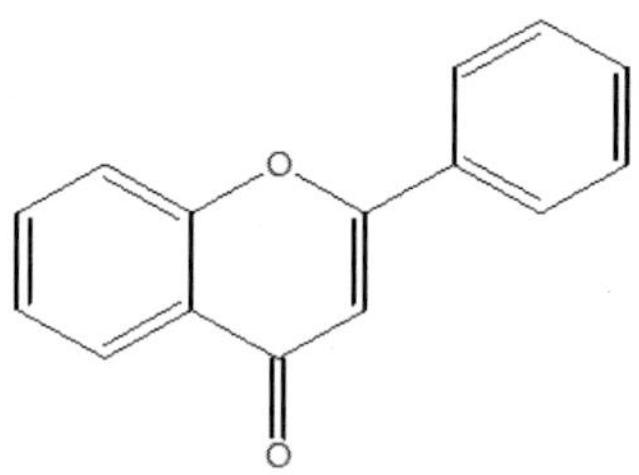

93

Figure 20. Chemical structures of amentoflavone **92** and flavone **93** from leaves of *L. monopetalum*.

Finally, we note that among the various kinds of natural compounds, the polyphenolic products constitute the main secondary metabolite in *Limoniastrum* genus. It is well known that polyphenolic constituents of various plants have been reported to exhibit several biological properties. In fact, these natural polyphenols are known as potent antioxidants and free radial scavengers and thus may exert a chemopreventive role towards cardiovascular and degenerative diseases [10, 87]. In addition, they have antibacterial, antiviral, anti-inflammatory and anti-carcinogenic potentials [88, 89].

CONCLUSION

Natural products from medicinal plants continue to play a fundamental role in drug discovery. Thus, the research for drug discovery from MAP is a multidisciplinary approach combining ethnopharmacology, botanical, biological, phytochemical, toxicological, pharmacological, and molecular methodology and techniques.

Plants of *Limoniastrum* genus and its sister *Limonium* from *Plumbaginaceae* family have been chemicaly investigated and almost 93 distinct compounds have been reported from theses genus. The *Limoniastrum* genus is represented in the the flora of Algeria by three species named: *Limoniastrum guyonianum, L. monopetalum and L. feei*, the last is endemic to Sahara and used in the traditional ethnopharmacopeae. The effectiveness in using *Limoniastrum* plant species against diverse ailments and to cure several illnesses in the Maghreb traditional medicine has scientifically been verified by the results obtained from biological activities (antimicrobial, antiparasitic, antiviral, antioxidant, anti-inflammatory, immunomodulatory, cytotoxic and antitumor) of some of their extracts as well as their variety of isolated compounds, especially the polyphenols derivatives.

It is important to mention that to the best of our knowledge, no work has been done on the pharmacokinetics studies of isolated active compounds from the genus. The knowledge of these action mechanisms will be helpful to understand their metabolism, their human health role and the possibility to obtain a new drug. Therefore, it is necessary to investigate the pharmacological activity, clinical, toxicity, structure-activity relationship, physiological pathways and mechanism action of isolated compounds from *Limoniastrum* genus, providing an opportunity for future research.

REFERENCES

[1] Cheriti, A., Belboukhari, M., Belboukhari, N. and Djeradi; H. (2012). “Phytochemical and biological studies on *Launaea* Cass. genus (Asteraceae) from Algerian Sahara”*Current Topics in Phytochem* 11, 67-80.

[2] Ibn al-Baytar, Dhiya al-Din. (1992). *Kitāb al-jāmi‘ li-mufradāt al-adwiya wa al-aghdhiya* [*Compendium on Simple Medicaments and Foods*], Ed. Dar Kotob Elmia, Libanon.

[3] Leclerc, L. (1877). *Traité des simples de ‘Abd Allāh ibn Aḥmad Ibn al-Bayṭār* [*Compendium on Simple of ‘Abd Allāh ibn Aḥmad Ibn al-Bayṭār*],Traduction, Impr. Nationale, Paris.

[4] Bellakhdar, J. (1997). *La pharmacopée marocaine traditionnelle. Médecine arabe ancienne et savoirs populaires* [*Moroccan traditional pharmacopoeia. Ancient Arabic medicine and popular knowledge*], IBIS Press.

[5] Leonti, M. and Verpoort, R. (2017). “Traditional Mediterranean and European herbal medicines” *J. Ethnopharmacol*. 199, 161-167.

[6] Petrovska, B.B. (2012). “Historical review of medicinal plants’ usage” *Pharmacogn. Rev*. 6, 1–5.

[7] Cheriti, A. and Belboukhari, N. (2015). Terpenoids of the Saharan medicinal plants *Launaea* Cass. Genus (Asteraceae) and their biological activities, in *Terpenoids and Squalene: Biosynthesis, Functions and Health Implications*, edited by Alanna B. Bates, 51-70, New York: Nova Science Publishers, USA.

[8] Rodríguez Villanueva, J., Esteban, J. M. and Rodríguez Villanueva, L. (2018). "Pharmacological Activities of Phytomedicines: A Challenge Horizon for Rational Knowledge" *Challenges* 9, 15-22.

[9] Gurib-Fakim, A. (2006). *Traditions of yesterday and drugs of tomorrow. Molecular Aspects of Medicine*. Special Edition, Elsevier Publications, UK.

[10] Ksouri, R., Meghdiche, W., Jalleli, I., Debez, A., Magné, C.; Hiroco, I. and Abdelly, C. (2012). "Medicinal halophytes: potent source of health promoting biomolecules with medical, nutraceutical and food applications." *Crit. Rev. Biotechno.* 32(4), 1–38.

[11] Rahmani, S. (2016). *Recherche des substances bioactives de la partie aérienne de la plante Limoniastrum feei* [*Research of bioactive substances of the aerial part of the plant Limoniastrum feei*], Doctorat Diss., UKM-Ouargla.

[12] Lefahal, M. (2014). *Etude phytochimique, biologique et activité anticorrosion de trois plantes médicinales Algériennes appartenant aux familles Plumbaginaceae, Tamaricaceae et Apiaceae* [*Phytochemical, biological and anticorrosion activity of three Algerian medicinal plants belonging to the families Plumbaginaceae, Tamaricaceae and Apiaceae*], Doctorat Diss., UM Constantine 1.

[13] Dobignard, A. and Chatelain, C. (2013). *Index synonymique de la flore d'afrique du nord* [*Synonymic index of the flora of North Africa*], Editions des conservatoire et jardin botaniques, Genève.

[14] Cheriti, A. (2000). *Medicinal Plants of the Bechar Region, Ethnopharmacologie studies*, CRSTRA Report, Algeria.

[15] Hernández-Ledesma, P., Berendsohn, W. G., Borsch, T., Mering, S. V., Akhani, H., Arias, S. and Uotila, P. (2015). "A taxonomic backbone for the global synthesis of species diversity in the angiosperm order Caryophyllales" *Willdenowia*, 45(3), 281–383.

[16] Koutroumpa, K., Theodoridis, S., Warren, B. H., Jiménez, A., Celep, F., Doğan, M., Romeiras, M. M.; Santos-Guerra, A.; Fernández-Palacios, J. M.; Caujapé-Castells, J.; Moura, M., Menezes de Sequeira, M. and Conti, E. (2018). "An expanded molecular phylogeny of *Plumbaginaceae*, with emphasis on *Limonium* (sea

lavenders): Taxonomic implications and biogeographic considerations" *Ecol. Evol.* 8(24): 12397–12424.

[17] Christenhusz, M. J. and Byng, J. W. (2016). "The number of known plants species in the world and its annual increase" *Phytotaxa*, 261(3), 201–217.

[18] Kubitzki, K. (1993). *Plumbaginaceae.* In *The Families and Genera of Vascular Plants, vol II* Edited by, Kubitzki, K.; Rohwer, J.G. and Bittrich, V. 523-530, Springer-Verlag, Berlin.

[19] Simpson, M.G. (2010). *Plant Systematic*, second edition, Academic Press.

[20] Dahlgren, R. (1983). "General aspects of angiosperm evolution and macrosystematics" *Nord. J. Bot.* 3, 119-149.

[21] APG. (2003). "An update of the Angiosperm Phylogeny Group Classification for the orders and families of flowering plants: APG II" *Botanical Journal of the Linnean Society* 141, 399–436. b) APG. (2016). "An update of the angiospermphylogeny group classification for the orders and families of flowering plants: APG IV" *Botanical Journal of the Linnean Society*,181(1), 1–20.

[22] Quezel, P. and Santa S. (1963). *Nouvelle flore de l'Algerie et des regions desrtiquesméridionales* [*New flora of Algeria and southern deserts regions*], CNRS, Paris.

[23] Ozenda, P. (2004). *Flore et Végétation du Sahara* [*Flora and vegetation of the Sahara*], CNRS, Paris.

[24] Maire, R. (1936)."Contributions a` l'étude de la Flore du l'Afrique du Nord" ["Contributions to the study of the flora of North Africa"] *Bull. Soc. Hist. Natu. Afr. Nord* 27, 241–270.

[25] Nieukerken, J.V.E. (2007). "Acalyptris Meyrick: Revision of the platani and staticis groups in Europe and the Mediterranean (Lepidoptera: Nepticulidae)" *Zootaxa* 1436, 1–48.

[26] Rahmani, S., Ziane, L., Belboukhari, N. and Cheriti, A. (2012). "The Saharan medicinal plant *Limoniastrum feei:* Ethnomedical survey and preliminary phytochemical screening of antibacterial extracts" *Phyto Chem & Bio Sub J.* 6 (2), 83-87.

[27] Krifa, M., Alhosin, M., Muller, C.D., Gies, J.P., Chekir-Ghedira, L., Ghedira, K., Mély, Y., Christian Bronner, C. and Mousli M. (2013). "*Limoniastrum guyonianum* aqueous gall extract induces apoptosis in human cervical cancer cells involving p16 INK4A re-expression related to UHRF1 and DNMT1 down-regulation" *J. Exp. Clin. Cancer Res.* 32 (30), 1-10.

[28] Krifa, M., Skandrani, I., Pizzi, A., Nasr, N., Ghedira, Z., Mustapha, N., Ghedira, K. and Chekir-Ghedira, L. (2014). "An aqueous extract of *Limoniastrum guyonianum* gall induces anti-tumor effects in melanoma injected mice via modulation of the immune response" *Food Chem. Toxicol.* 69, 76–85.

[29] Trabelsi, N., Oueslati, S., Ksouri, R., Nassra, M., Marchal, A., Krisa, S., Abdelly, C., Merillon, J. M. and Waffo-Teguo, P. (2014). "The antioxidant properties of new dimer and two monomers of phenolic acid amides isolated from *Limoniastrum guyonianum*" *Food Chem.* 146, 466–471.

[30] Cheriti, A., Belboukhari, N. and Hacini, S. (2013). "Ethnopharmacological survey and phytochemical screening of some medicinal Asteraceae from Algerian Sahara" *Phytochem & BioSub J.* 7 (2), 52-56.

[31] Belboukhari, N. and Cheriti, A. (2005). "Anti-microbial activity of aerial part crude extracts from *Limoniastrum feei*" *Asian J. Plant Sci.* 4, 496–498.

[32] Hammami, S., Nguir, A., Saidana, D., Cheriaa,J. and Mighri; Z. (2011). "Chemical analysis and antimicrobial effects of essential oil from *Limoniastrum guyonianum* growing in Tunisia" *J. Med. Plants Res.* 5, 2540–2545.

[33] Salah Ramadan, B., Cheriti, A., Belboukhari, N. and Zaouani, M. (2008). "Evaluation de la toxicité de deux Asteraceae du Sahara Algérien: *Launaea Arbonescens* et *Bubonium gravoelens*" [""Evaluation of the toxicity of two Asteraceae from the Algerian Sahara: Launaea Arbonescens and Bubonium gravoelens"]. *Annales Univ Bechar* 4, 48-54.

[34] Boulenouar, N., Marouf, A., Cheriti, A. and Belboukhari, N. (2012). "Medicinal Plants Extracts as Source of Antifungal Agents against *Fusarium oxysporum* f. sp. *albedinis*" *J. Agr. Sci. Tech.* 14, 659-569.

[35] Dahane-Rouissat, L., Cheriti, A., Marouf, A., Kandappa H. R. and Govender, P. (2015). "Molluscicidal activity of the Saharian medicinal plants *Limoniastrum feei* and *Launaea nudicaulis* against the fresh water snail *Lymnaea stagnalis*" *J. Biodiv. Env. Sci.* 7(6), 242-248.

[36] Bourmita, Y., Cheriti, A., Ould El Hadj, M. D., Mahmoudi, K. and Belboukhari, N. (2013). "Anti-termitic Activity of Aqueous Extracts from Saharan Toxic Plants against *Anacanthotermes ochraceus*" *J. Entomology* 10(4), 207-213.

[37] Bourmita, Y. (2014). *Toxicité comparée des extraits de quelques plantes spontanées de la région de Béchar chez des termines de type saharien* [*Comparative toxicity of extracts of some spontaneous plants from the Bechar region in Saharan-type termines*], Doctorat Diss., UKM-Ouargla.

[38] Halliwell, B. and Gutteridge, J.M. (1990) "The antioxidants of human extracellular fluids" *Arch. Biochem. Biophys.* 280, 1–8.

[39] Mut-Salud, N., Álvarez, P.J., Garrido, J.M., Carrasco, E., Aránega, A. and Rodríguez-Serrano, F. (2016). "Antioxidantintake and antitumor therapy: Toward nutritional recommendations for optimal results" *Oxid. Med. Cell. Longev.* 6719534, 1-19.

[40] Keffous, F., Belboukhari, N., Sekkoum, K., Djeradi, H., Cheriti, A. and Aboul-Enein, H. Y. (2016). "Determination of the antioxidant activity of *Limoniastrum feei* aqueous extract by chemical and electrochemical methods" *Cogent Chemistry* 2(1), 1186141, 1-11.

[41] Bettaib, J., Talarmin, H., Zar Kalai, F., Giroux-Metges, M. G. A. and Ksouri, K. (2017). "Limoniastrum guyonianum prevents H_2O_2-induced oxidative damage in IEC-6 cells by enhancing enzyamtic defense, reducing glutathione depletion and JNK phosphorylation" *Biomed. Pharmaco.* 95, 1404-1411.

[42] Debouba, M., Zouari, S. and Zouari, N. (2013). "Evaluation of Antioxidant Status of Two *Limoniastrum* Species Growing Wild in Tunisian Salty Lands" *Antioxidants* 2, 122-131.

[43] Krifa, M., Bouhlel, I., Skandrani, I., Chekir-Ghedira, L. and Ghedira, K. (2014). "Antioxidant, mutagenic and antimutagenic activities of an aqueous extract of *Limoniastrum guyonianum* gall" *Drug Chem. Toxicol.* 37(1), 76–82.

[44] Krifa, M., Pizzi, A., Mousli, M., Chekir-Ghedira, L., Leloup, L. and Ghedira, K. (2014). "*Limoniastrum guyonianum* induces apoptosis in colorectal cancer cells by inhibiting calpain activity" *Tumor Biol.*35(8), 7877-85.

[45] Krifa, M., Bouhlel, I., Chekir-Ghedira, L. and Ghedira, K. (2013). "Immunomodulatory and cellular anti-oxidant activities of an aqueous extract of *Limoniastrum guyonianum* gall" *J. Ethnopharm.* 146, 243–249.

[46] Lemarie, F., Chang, C.W., Blatchford, D.R., Amor, R., Norris, G., Tetley, L., McConnell, G. and Dufes, C. (2013). "Antitumor activity of the tea polyphenol epigallocatechin-3-gallate encapsulated in targeted vesicles after intravenous administration" *Nanomedicine* 8, 181–192.

[47] Boudermine, S., Malafronte, N., Bghidja, N., Benayech, S., Esposito, A., Velotto, S. and Severino, L. (2014). "Phytochemical investigation and antiproliferative activity of *Limoniastrum feei* (girard) batt." *Pharmacology on Line* 3, 141–144.

[48] Medzhitov, R. (2010). "Inflammation 2010: new adventures of an old flame" *Cell.* 140, 771–776.

[49] Chen, L., Deng, H., Cui, H., Fang, J., Zuo, Z., Deng, J., Li, Y., Wang, X. and Zhao, L. (2018). "Inflammatory responses and inflammation-associated diseases in organs" *Oncotarget* 9(6), 7204-7218.

[50] He, G. and Karin, M. (2011). "NF-κB, STAT3 - key players in liver inflammation and cancer" *Cell Res.* 21, 159–168.

[51] Swanson, H. (2016). *Flavonoids, inflammation and cancer* Ed World Scientific Pub.

[52] Allaoui, A., Belboukhari, N., Sekkoum, K., Mat Jais, A.M., Seddiki, L. and Cheriti, A. (2017). "*In Vivo* analgesic activity of aqueous extract from leaves of *Limoniastrum feei*" *MOJ Bioequiv Availab.* 4(3), 282–283.

[53] Rahmani, S., Belboukhari, N., Sekkoum, K. and Cheriti, A. (2016). "Evaluation de l'activité anti-inflammatoire d'extraits aqueux de feuilles *Limoniastrum feei* (plumbaginacea)" ["Evaluation of the anti-inflammatory activity of aqueous leaf extracts Limoniastrum feei (plumbaginacea)"]. *Algerian J. Arid Envir.* 6(1), 80-86.

[54] Kennedy, D.O. and Wightman, E.L. (2011). "Herbal extracts and phytochemicals: Plant secondary metabolites and the enhancement of human brain function" *Advances in Nutrition* 2, 32-50.

[55] Newman, D.J. and Cragg, G.M. (2012)."Natural products as sources of new drugsover the 30 years from 1981 to 2010" *J. Nat. Prod.* 75, 311–355.

[56] Ulrich-Merzenich, G., Zeitler, H., Jobst, D., Panek, D., Vetter, H. and Wagner, H. (2007). "Application of the "-Omic-" technologies in phytomedicine" *Phytomedicine* 14, 70–82.

[57] Messaoudi, R., Cheriti, A. and Bourmita, Y. (2018). "Bioassay-guided isolation of the major compound with antioxidant activity from the algerian medicinal plant *bubonium graveolen*" *Asian J. Pharm. Clin. Res.* 11(11), 424-426.

[58] Lledo, M. D., Crespo, M. B., Cox, A. V., Fay, M. F. and Chase, M. W. (2000). "Polyphyly of *Limoniastrum* (*Plumbaginaceae*): evidence from DNA sequences of plastid *rbc*L, *trn*L intron and *trn*L-F intergene spacer" *Botan. J. Linnean Soc.* 132, 175–191.

[59] Lledo, M. D., Crespo, M. B., Cameron, K.M., Fay, M. F. and Chase, M. W. (1998). "Systematics of *Plumbaginaceae* based upon cladistic analysis of *rbc*L sequence data" *Systematic Botany* 23, 21–29.

[60] Ye, G. and Huang, C. (2006). "Flavonoids of *Limonium aureum*" *Chem. Nat. Comp.* 42(2), 232-234.

[61] Benaissa, O., Amrani, A., Bicha, S., Zama, D., Benayache, F., Marchioni, E. and Benayache, S. (2013). "Free radical scavenging

action of phenolic compounds from *Limonium bonduelli* (*Plumbaginaceae*)" *Pharm. Lett.* 5 (5), 234-240.

[62] Lie-Chwen, L. and Cheng-Jen, C. (2000). "Flavonoids and Phenolics from *Limonium sinense*" *Planta Med.* 66(4), 382-383.

[63] Medini, F., Bourgou, S., Lalancette K. G., Snoussi, M., Mkadmini, K., Coté, I., Abdelly, C., Legault, J. and Ksouri, R. (2015). "Phytochemical analysis, antioxidant, anti-inflammatory, and anticancer activities of the halophyte *Limonium densiflorum* extracts on human cell lines and murine macrophages" *South Afr. J. Bot.* 99, 158-164.

[64] Korul Kina, L.M., Shul'ts, E. E., Zhusupova, G.E., Abilov, Zh. A., Erzhano, K.B. and Chaudri, M.I. (2004). "Biologically Actives Compounds From *Limonium Gmelinii* and *L. Popovii*" *Chem. Nat. Comp.* 40(5), 465-471.

[65] Kozhamkulova, Z.A., Radwan, M.M., Zhusupova, G.E., Abilov, Z. Zh., Rahadilova, S.N. and Ross, S.A. (2010). "Gmelinoside I, a new flavonolglycoside from *Limonium gmelinii*"*Nat.Prod. Commun.* 5(7), 1061-1062.

[66] Gadetskaya, A.V., Zhusupova, G.E., Tarawneh, A.H., Ma, G., Cutler, S.J. and Ross, S.A. (2015). "Therapeutic efficacy of phenolic compounds from *Limonium* in acute lymphoblastic leukemia"*Planta Med.* 81, PC3.

[67] Asen, S. and Plimmer, J.R.S. (1972). "4,6,4'-Trihydroxyaurone And Other Flavonoids From *Limonium*" *Phytochemistry*11, 2601-2603.

[68] Ahmed, K.M., Kandil, F.E. and Mabry, T.J. (1999). "Anticancer tannin and other phenolics from *Limonium axillare* (*Plumbaginaceae*)" *Asian J. Chem.* 11, 261-263.

[69] Iwashina, T. (2013). "Flavonoid Properties of five Families newly Incorporated into the Order Caryophyllales (Review)"*Bull. Natl. Mus. Nat. Sci., Ser. B* 39(1), 25–51.

[70] Lee, J.I., Kong, C.S., Jung, M. E., Hong, J. W., Lim, S. Y. and Seo, Y. (2011). " Antioxidant Activity of the Hlaphyte *Limonium tetragonum* and its Major Active Components". *Biothecno. Bioproc. Eng.*16, 992-999.

[71] Kandil, F.E., Ahmed, K.M., Hussieny, H.A. and Souliman, A.M. (2000). "A new flavonoid from *Limoniumaxillare*"*Archivder Pharmazie: Pharmaceutical and Medicinal Chemistry*, 333, 275-277.
[72] Ross, S.A. (1984). " Myricetin-3'-MethylEther-7-Glucoside from *Limonium Sinuatum*" *J. Nat. Prod.* 47(5), 862-364.
[73] Asen, S., Norris, K. H., Stewart, R. N. and Semeniuk, P. (1973). "Effect of pH, anthocyanin, and flavonoid co-pigments on the color of statice flowers" *J. Am. Soc. Horticult. Sci.* 98, 174–176.
[74] Bachir, A.K., Abdalla, A.A., Wasfi, I.A., Hassen, E.S., Amir, M.H. and Crabb, T.A. (1994). "Flavonoids from *Limonium axillare*" *Int. J. Pharmacognos* 32, 366-372.
[75] Zhusupova, G.E. and Abil'kaeva, S.A. (2006). "Flavanes from *Limonium gmelinii*" *Chem. Nat. Comp.* 42(1), 112-113.
[76] Kuo, Y.C., Lin, L.C., Tsai, W. J., Chou, C. J., Kung, S.H. and Ho, Y. H. (2002). "Samarangenin B from *Limoniumsinense* Suppresses Herpes Simplex Virus Type1 Replication in Vero Cells by Regulation of Viral Macromolecular Synthesis" *Antimicrobial Agents and Chemotherapy* 46(9), 2854– 2864.
[77] Saıdana, D., Mahjoub, S., Boussaada, O., Chriaa, J., Mahjoub, J., Chéraif, I., Daami, M., Mighri, Z. and Helal, A.N. (2008). "Antibacterial and Antifungal Activities of the Essential Oils of Two Saltcedar Species from Tunisia" *J. Am. Oil Chem. Soc.* 85, 817–826.
[78] Chaabi, M., Beghidja, N., Benayache, S. and Lobstein, A. (2008). "Activity-guided isolation of antioxidant principles from *Limoniastrum feei* (Girard) Batt." *Z Naturforsch C* 63, 801–807.
[79] Belboukhari, N. and Cheriti, A. (2007). "Flavonoids of *Limoniastrum feei*" *Res. J. Phytochem.* 1(2), 74–78.
[80] Rahmani, S., Belboukhari, N., Cheriti, A., Bourmita, Y. and Ould El Hadj, M.D. (2013). "Glucoside flavonoids from the antimicriobial extract of the Saharan medicinal plant *Limoniastrum feei" Biochemistry Ind J.* 7(4), 137-140.
[81] Belboukhari, N. and Cheriti, A. (2009). "Analysis and isolation of saponins from *Limoniastrum feei* by LC-UV" *Chem. Nat. Comp.* 45, 756–758.

[82] Rahmani, S., Belboukhari, N. and Cheriti, A. (2014). "Phytochemical investigation of bioactive extract from endemic medicinal plant *Limoniastrum feei* (Girard) Batt (*Plumbaginaceae*)" *Asian J. Chem.* 26, 365–368.

[83] Allaoui A., Rahmani, S., Belboukhari, N., Cheikh, N., Cheriti, A. and Sekkoum, K. (2015). "Isoflavone and flavone derivatives from aerial part of *Limoniastrum feei*" *Der Pharma Chemica*, 7(8):17-20.

[84] Cheriti, A. (2018). "Phytochemistry and biological activities of some endemic Saharan medicinal plants" *Paper presented at the International Days of Organic Chemistry*, Annaba, December 3-5.

[85] Trabelsi, N., Megdiche, W., Ksouri, R., Falleh, H., Oueslati, S., Bourgou, S., Hajlaoui, H. and Abdelly, C. (2010). "Solvent effects on phenolic contents and biological activities of the halophyte *Limoniastrum monopetalum* leaves" *LWT - Food Sci.Techno.* 43, 632–639.

[86] Trabelsi, N., Oueslati, S., Falleh, H., Wafoo-Teguo, P., Yorgos, P., Merillon, J. M., Abdelly, C. and Ksouri, R. (2012). "Isolation of powerful antioxidants from the medicinal halophyte *Limoniastrum guyonianum*". *Food Chem.* 135, 1419–1424.

[87] Djeradi, H., Rahmouni, A. and Cheriti, A. (2014). "Antioxidant activity of flavonoids: A QSAR modeling using Fukui indices descriptors" *J. Mol. Model.* 20, 2476-83.

[88] Chia-Jui, W. and Gow-Chin, Y. (2012). "Chemopreventive effects of dietary phytochemicals against cancer invasionand metastasis: phenolic acids, monophenol, polyphenol, and their derivatives" *Cancer Treatment Rev.* 38, 76–87.

[89] Djeridane, A., Yousfi, M., Nadjemi, B., Boutassouna, D., Stocker, P. and Vidal, N. (2006). "Antioxidant activity of some Algerian medicinal plants extracts containing phenolic compounds" *Food Chem.* 97, 654–660.

INDEX

A

B

C

D

E

F

I

J

K

L

M

Q

R

S

T

U

V

W

Y

Z

α

Related Nova Publications

Occurrences, Structure, Biosynthesis, and Health Benefits Based on Their Evidences of Medicinal Phytochemicals in Vegetables and Fruits. Volume 11

Editor: Noboru Motohashi, Ph.D.

Series: Food and Beverage Consumption and Health

Book Description: These chapters will provide more advanced information to researches in studying new drug designs of phytochemicals.

Hardcover ISBN: 978-1-53614-071-2
Retail Price: $195

Occurrences, Structure, Biosynthesis, and Health Benefits Based on Their Evidences of Medicinal Phytochemicals in Vegetables and Fruits. Volume 10

Editor: Noboru Motohashi, Ph.D.

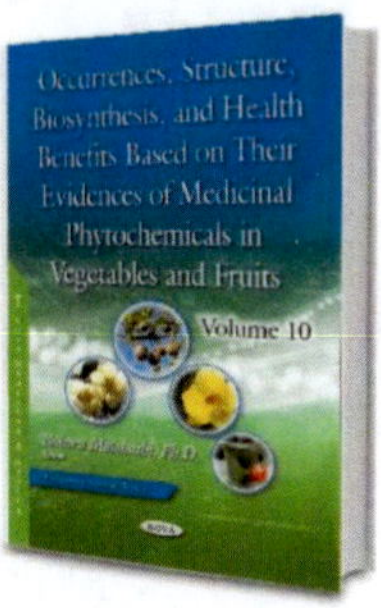

Series: Food and Beverage Consumption and Health

Book Description: In East Asia, Korean ginseng has long been used as a private medicine and nourishing tonic, used for maintaining physical strength as well as treating and preventing various diseases.

Hardcover ISBN: 978-1-53614-141-2
Retail Price: $160

To see complete list of Nova publications, please visit our website at www.novapublishers.com